Automation and Robotics in Agriculture

Automation and Robotics in Agriculture

Hazem Shawky Fouda

Delve
PUBLISHING

www.delvepublishing.com

Automation and Robotics in Agriculture

Hazem Shawky Fouda

Delve Publishing
224 Shoreacres Road
Burlington, ON L7L 2H2
Canada
www.delvepublishing.com
Tel: 001-289-291-7705
001-905-616-2116
Fax: 001-289-291-7601
Email: orders@arclereducation.com

ISBN: 978-1-77407-648-4 (Hardcover)

ABOUT THE AUTHOR

Hazem Shawky Fouda has a PhD. In Agriculture Sciences, obtained his PhD. from the Faculty of Agriculture, Alexandria University, 2008, MSc. In Agriculture Sciences from the Faculty of Agriculture, Alexandria University in 2004, Post-Grade Diploma in Cotton, 2001, BSc. in Agriculture Sciences, from the Faculty of Agriculture, Alexandria University, 1997, He is working in Cotton Arbitration & Testing General Organization (CATGO) since 1999 till present. Was working in the International Cotton Training Center (ICTC) – Cotton Arbitration & Testing General Organization (CATGO) from 2000 till 2015, as a Lecturer & Classer's Trainer for Egyptian and foreigner classers, technicians, ginners, spinners & traders in all cotton aspects starting from picking seed cotton till preparing bales for export including other cotton aspects and training them practically on classing cotton. Besides that he was an editor and active member in the Research & Translation Committee, participating in issuing weekly, monthly and annually issues about the international & local cotton market including price trends and direction, recent developments & researches concerning cotton production, protection, harvesting, ginning, fiber testing, spinning & weaving since its foundation in 2000 till 2014 and since 2015 till 2018 he worked as an inspector from 2018 till present works as a consultant..

TABLE OF CONTENTS

LIST OF FIGURES

LIST OF TABLES

LIST OF ABBREVIATIONS

3-D	three-dimensional
AHRS	attitude heading reference system
ANN	artificial neural network
ANNHS	artificial neural network-harmony search
API	autonomous plant inspection
B	boron
BCD	binary code to decimal
BPNN	back propagation neural network
BSM	blurred form model
Ca	calcium
CAN	controller area network
CBIR	content-based image retrieval system
CCD	charge-coupled device
CCF	comprehensive color feature
CCM	co-occurrence method
CHT	circular Hough transform
CLAHE	contrast limited adaptive histogram equalization
CNN	convolutional neural network
CPU	central processing unit
C-V	Chan-Vese
DE	delta E
DIAS	Danish Institute of Agricultural Sciences
DIO	digital input/output
DLL	dynamic-link libraries
DOF	degree-of-freedom
E/H	electrohydraulic
ERS	entropy rate superpixel
FAST	features from an accelerated segment test

Fe	iron
GA	genetic algorithm
GIS	geographical information systems
GLCM	gray-level co-occurrence matrix
GP	generalized Pareto
GPS	global positioning system
GPU	graphical processing unit
HMS	headland management systems
HOG	histogram of oriented gradients
HRI	human-robot interaction
HSeg	hue plane threshold segmentation
HSL	hue saturation luminance
HSV	hue, saturation, value
IA	immune algorithm
IEEE	Institute of Electrical and Electronics Engineers
IFR	international federation of robotics
IFToMM	international federation for the theory of machines and mechanisms
IMU	inertial measurement unit
INS	inertial navigation system
IPS	integrated pond system
ISO	International Standard Organization
JAUGS	Joint Architecture for Unmanned Ground Systems
K	potassium
k-NN	k-nearest neighbors
LAN	local-area network
LBP	local binary pattern
LGGP	local gray Gabor pattern
MEMS	micro-electromechanical systems
MLP	multilayer perceptron
M-SVM	multi-class support vector machine
NIR	near-infrared
ORB	oriented fast and rotated brief
Panicle-SEG	panicle segmentation
PC	personal computer

PCA	principal component analysis
PCNN	pulse coupled neural networks
PCR	polymerase chain reaction
PF	particle filter
PLC	programmable logic controller
PTO	power takeoff
PTP	point-to-point
RAS	recirculating aquaculture systems
RGB	red-green-blue
ROI	region of interest
RTK-DGPS	real-time kinematic differential global positioning system
RTK-GPS	real-time kinematic global positioning system
SCARA	selective compliant arm for robot assembly
SCARA	selective compliant robot arm
SCC	somatic cell count
SIFT	scale-invariant feature transform
SLIC	simple linear iterative clustering
SS	speed-sprayer
STTPS	single-truss tomato production systems
SVM	support vector machine
USB	universal serial bus
UV	ultraviolet
Wi-Fi	wireless fidelity

PREFACE

Over the last few decades, wide-ranging research has been carried out on the uses of automation and agricultural robots to several greenhouse and field operations. Moreover, the technical basics and their feasibility reports have also been broadly demonstrated. Adverse interference, unstructured environment, and diversified operational processes are the key factors blocking the commercialization of agricultural robotic operations. However, the development of novel automation techniques, information techniques, and smart sensors has resulted in considerable success in agricultural robots in recent years. This book fundamentally intends to offer the reader a detailed overview of the contemporary state of the art in automation and agricultural robots.

Recent progressions in mechanized farming technology have radically transformed the current agricultural industry to fulfill the rising and incessantly changing needs of consumers around the globe. To ensure the sufficient supply and production of quality crop plants, agriculture robots have appeared as a potential solution.

In forestry and agriculture, robotics has played a substantial role. Farmers are aware of their necessity for automatic guidance of vehicles to minimize the harm to the growing zones of the soil. Automatic handling, processing, and sensing of agricultural products are now common, while there are considerable mechanization and instrumentation of livestock processes. The blend of machine vision with GPS (global positioning by satellite) permits a tractor to track a row of the crop, executing a promontory turn at the row end. Although automation and robotics are making rapid inroads into different areas of agricultural farming, there is a great demand for improvement of existing robotic technology and the development of novel techniques.

This book contains eight chapters. The first two chapters of the book introduce the audience with fundamental concepts of automation, robotics, and agricultural automation. Chapter 3 deals with robotic technology and its applications in agricultural systems. Recently, machine vision systems have become popular for the detection of objects. Chapter 4 specifically discusses the applications of machine vision systems in crop farming.

Chapter 5 focuses on the fundamental concepts of actuation mechanisms and their applications in agricultural systems. Chapter 6 discusses the science of human-robot interactions in agricultural environments. The advent of modern robotic technology has revolutionized the agricultural mechanism. Chapter 7 illustrates the use of robotic technology for automatic spraying in the fields. Finally, Chapter 8 discusses the economic and commercial aspects of agricultural robots and their diverse applications.

This book encompasses all of the fundamental aspects of automation and robotic technology used in agriculture. Therefore, this book can be considered as reference material for readers from diverse backgrounds including forestry, agriculture, robotics, and automobiles industry.

—Author

Chapter 1

Fundamentals of Automation and Robotics

CONTENTS

1.1. INTRODUCTION

Robots could be considered as the utmost advanced automatic structures and robotics, as a method and scientific discipline could be considered as the progression of automation with interdisciplinary incorporation with other technological fields (Chapman, 2003; Chui et al., 2015). An automatic system could be stated as a system that is capable to repeat precise operations usually with a low degree of manipulative and intellectual levels, however, that could be simply programmed in agreement by demands of output (Keramas et al., 1998; Xie, 2003).

It is important to note that an automatic structure is usually capable to perform one action for which its mechanical system had been designed. Features of flexibility rely on the possibility of reprogramming the governing unit, which is usually capable only to alter the time sequence of the intended operations. So, an automatic structure of industrial kind could be thought of as composed of three parts (Mooring et al., 1991; Slaughter, ing 1997; Kolluru et al., 2000):

- Hardware with electrical, pneumatic, mechanical and hydraulic components that deliver the mechanical capability to execute an *a priori*-determined operation of manipulation and/or movement;
- Operation and control counterpart with electronic parts and software that gives the capability of independence and flexibility to the working of the system. The two components are essential in an automatic structure and are incorporated in the sense that their operation and design must be considered as an exclusive goal to acquire and operate an automatic structure with the best shows. Thus, an automatic structure in which the hardware component is preponderant cannot have appropriate flexibility for flexible manufacturing in agreement by the demands of output and market. In certain cases, these restricted solutions are still needed from productivity, when the product of a huge production is engrossed from the market with certain regularity throughout a period lengthier than the repayment time of the automatic structure. Those systems with less flexibility are usually denominated as 'rigid automatic structures.'
- Automatic structures are frequently designed to have the possibility of simple modernizing of their structure and operation to be adjusted more rapidly to the variable demands of production

and design of products. These flexible structures are denominated as 'flexible automatic structures' (Dietrich et al., 2009).

Besides, it is important to note that a structure with merely the mechanical counterpart, though versatile, cannot be deliberated an automatic structure as the updating of its operation is not gone through means of control units. Such a structure could be appropriately reprogrammed, but it needs the physical action of one or more human operatives to alter some parts of the machinery or their running. Thus, these structures could be properly named as mechanical systems or machinery (Angeles and Angeles, 2002; Todd, 2012) (Figure 1.1).

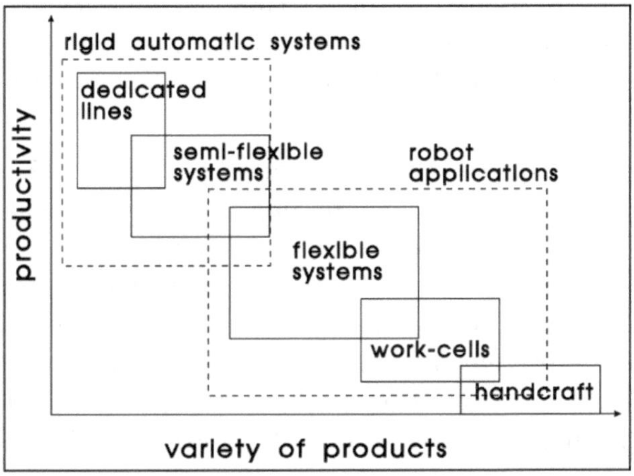

Figure 1.1*:* Variety of robotized and automatic systems as functions of product demand and productivity level.

Source: https://link.springer.com/chapter/10.1007%2F978-1-4020-2110-7_1.

Likewise, control structures can't be considered as automatic structures, as they are not capable to perform mechanical actions, though they could be provided with a high level of flexibility related to memory capability and re-programmability. Thus these structures could be appropriately named as electrical-electronic-informatics systems or control units (Elattar, 2008; Poole, 2012; Li and Ge, 2013).

Though, an automatic structure is designed and built by utilizing a suitable integration and combination of control units and mechanical systems. Certainly, the accomplishment and engineering application of automatic structures strongly rely on the practical incorporation of the aforesaid counterparts (Paulson, Jr. 1985; Skibniewski and Hendrickson, 1990; Richard, 2005).

1.2. AUTOMATIC SYSTEMS AND ROBOTS

A robot can be defined as a structure that is capable to perform numerous manipulative tasks with tools, objects, and even its extremity (end-effector) with the ability to be reprogrammed for numerous kinds of operations. There is an incorporation of control and mechanical counterparts; however, it even contains additional components and equipment, concerned with artificial intelligence and sensorial capabilities. Thus, the simultaneous design and operation integration of all the aforesaid systems will deliver a robotic system, as shown in Figure 1.2 (Everett and Slocum, 1994; Daerden et al., 2001).

In actuality, more than in automatic structures, robots can be characterized as having concurrent lyre-programming and mechanical capabilities. The mechanical ability is concerned with multipurpose characteristics in manipulative tasks because of the mechanical counterparts, and re-programming abilities concerned with flexible features in control abilities because of the electric-electronics-informatics counterparts. Thus, a robot can be considered as a complicated system that is composed of numerous devices and systems to give (Montemerlo, 2001; Saunders, 2004; Grift et al., 2008):

- Intellectual capabilities (for decision, memory, and control);
- Sensorial capabilities (like to human beings and/or precise others); and
- Mechanical capabilities (force and motion).

Initially, industrial robots were established to assist industrial procedures through substituting human operators in hazardous and repetitive operations, and harmful environments. Today, additional needs encourage further utilization of robots, even from clean technical perspectives, like productivity enhancement and improvements in product quality. Therefore, the first robots have been progressed to complex structures with further capabilities (Bard, 1986; Cusack, 1994; Balaguer et al., 2008).

However, referring to Figure 1.2, an industrial robot can be assumed as composed of: a manipulator arm or mechanical system (mechanical structure), whose purpose comprises of performing the manipulative operation and/

or collaborations with the environment; sensorial equipment (external and internal sensors) that is outside or inside the mechanical system, and whose purpose is to get information on the robot state and situation, which is in the robot range (Karwowski et al., 1998; Hasegawa, 2006; Zavadskas, 2010);

- A controller (control unit), which gives an explanation of the information from the sensorial equipment for controlling of the complete systems and provides the actuation signals for the operation of the robot and completing of desired tasks;
- A power unit, which gives the desired energy for the system and its appropriate transformation in magnitude and nature as needed for the robot components;-computer facilities, which are needed to expand the computation ability of the control unit and even to give the ability of artificial intelligence.

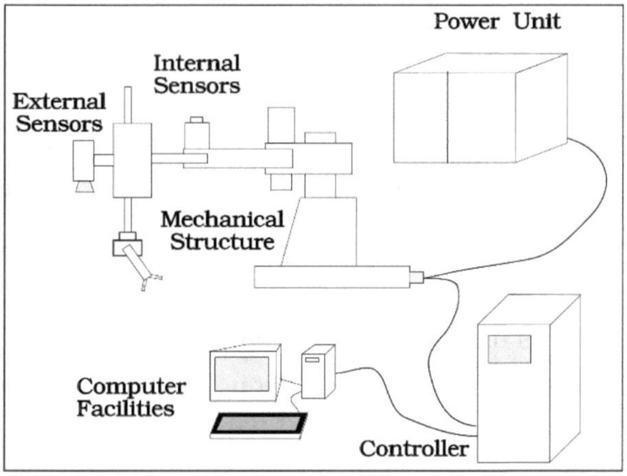

Figure 1.2: Constituents of an industrial robot.

Source: https://link.springer.com/chapter/10.1007%2F978-1-4020-2110-7_1.

Therefore, the aforesaid combination of sub-systems provides the three major simultaneous approaches to a robot, i.e., mechanical action, data amplification, and reprogram ability (Elkmann et al., 2002; Zhang et al., 2003).

Subsequently, the fundamental ability of robotic structures can be recognized (Arimoto et al., 1984; Heinzmann and Zelinsky, 1999):

- Mechanical versatility;
- Re-programmability.

Mechanical versatility of a robot could be understood as the ability to perform a diversity of tasks due to the mechanical and kinematic design of its manipulator's arm. Re-programmability of a robot could be understood as the litheness to perform a diversity of task operations due to the ability of computer facilities and its controller. These basic demonstrations give appropriate flexibility for the accomplishment of numerous different tasks in the same or better way than human arms. Currently, robots are well-established parts in industrial automation as they replaced human operators in operations and situations, which are (Steels, 1995; Biggs and MacDonald, 2003):

- Repetitive: with low-level technical and cultural content;
- Difficult: for human operators;
- Dangerous: for unhealthy environments and/or manipulative tasks; and
- Tiresome: for manipulative tasks demanding greater energy than that delivered by a human operator.

Besides, the utilization of robots is well-established in industrial manufacturing as robots are (Smithers, 1997; Song et al., 2011):

- Valuable for limited or unhealthy environments, which could be unfeasible or dangerous for human operators; and
- Versatile, as they can operate in diverse situations.

Besides the aforementioned motivations of the technical side, robots, and robotic structures are presently used with the purpose of (Kamimura et al., 2003; 2005):

- Improving product quality, in terms of improved manufacturing and constant characteristics;
- Growing productivity, with an enhancement of the operation velocity;
- Attaining flexible production as an ability of quick adjustment of production to required alterations in manufacturing; and
- Reducing production costs, which can be associated with better utilization of the machinery;
- The mechanical ability of a robot is because of the mechanical sub-system that usually is identified and denominated as the 'manipulator,' as its purpose is the manipulative task.

The term manipulation mentions numerous operations, which comprise (Hornby et al., 2001; Barone et al., 2004):

- Releasing and grasping of objects; and
- Interaction with the surroundings and/or with objects not associated with the robot; transportation and movement of objects and/or robot extremity.

Thus, the mechanical sub-system provides mechanical versatility to a robot by dynamic and kinematic capability throughout its operation. Manipulators can be categorized according to the kinematic chain of their architectures as (Suh et al., 1991; Thomas and Wahl, 2001):

- Parallel manipulators, when they can be demonstrated as closed kinematic chains in which the links are linked to each other hence that polygonal gaps can be determined; and
- Serial manipulators when they can be demonstrated as open kinematic chains in which the links are linked successively through binary joints;

Further, the kinematic chains of manipulators could be spatial or planar relying on which space they work. Mostly industrial robotic manipulators are of the sequential type, though lately parallel manipulators had stimulated great interest and are even utilized in industrial applications.

In general, to implement the same manipulative tasks as human operators, a manipulator is made of the following mechanical subsystems:

- A wrist, whose purpose is to orientate the extremity;
- An arm, which is dedicated for the execution of large movements, mostly as translations; and
- An end-effector (the component of the robot interacting with the environment), which is the exploiter extremity that interrelates with the environment.

Numerous different architectures had been designed for every of the aforementioned manipulator sub-systems as a function of necessary precise capabilities and characteristics of precise mechanical designs. It is important to note that though the mechanical design of a manipulator is founded on common mechanical parts, like all types of transmissions, the distinctiveness of robot design and operation needs advanced design of those parts in terms of dimensions, designs, and materials, due to the need for extreme compactness, reliability, and lightness.

The sensing ability of a robot is got by utilizing sensors appropriate for knowing the status of the surrounding environment and robot itself. The sensors for robot status are of main importance as they permit the control of the operation of the manipulator. Thus, they are generally installed on the manipulator itself to monitor the basic features of manipulations, like velocity, force, and position. Furthermore, an industrial robot can be equipped with precise and/or innovative sensors, which provide better or human-like sensing capability. Thus, a large variety of sensors can be utilized, to which the reader is recommended to refer to in precise literature. The control unit is of vital importance since it provides the capability for the intelligent and autonomous operation to the robot and it executes the following purposes (Meystel, 1991):

- Elaboration and acquisition of sensor signals from the manipulator and nearby environment;
- The ability of memory and computation, which is required for the aforementioned aim and robot re-programmability; and
- Regulation of the manipulator wave as a function of desired and present values of main dynamic and kinematic variables using appropriate computations and programming.

In specific, an intellectual ability has been added to certain robotic structures concerned generally with decision ability and memory of previous experiences by utilizing the techniques and means of artificial intelligence and expert systems. Yet, most of the existing industrial robots have no intelligent capability as the control unit accurately operates for the assigned tasks within industrial environments. The control systems that are utilized in robots can be categorized as:

- Units with pneumatic logic;
- PLC (Programmable Logic Controller) units;
- Electro-mechanical sequencer units with end-stroke stops;
- Electronic units with logic;
- Minicomputers;
- Computers; and
- Micro-processors.

Indeed, these days industrial robots are generally equipped with minicomputers, as the development of low-cost PCs has determined the extensive utilization of PCs in robotics so that sequencers, which are going to be limited to PLC units only, will be utilized mainly in low-flexible or rigid automation systems.

As robots could be considered complicated systems both in operation and design even inside work-cells, it is of vital importance and eases to have the prospect to simulate robot operation and even the complete robotic systems. Utilizing experimental tests and/or numerical methods even with laboratory models can simulate robots. Thus, an expert in robotics and automation should be attentive and be capable to work with methodologies and techniques which are beneficial in running and designing the aforementioned simulations (Koza and Rice, 1992; Wu and Zhou, 2007).

1.3. EVOLUTION AND APPLICATIONS OF ROBOTS

It is well recognized that the word 'robot' was created by Karel Capek in 1921 for a theater play assigning with cybernetic workers, who substitute humans in heavy work. Certainly, even nowadays the word 'robot' had an extensive meaning that comprises any system that can work independently for a specified class of tasks. Occasionally intelligent capability is involved as a fundamental property of a robot, as revealed in many works of movies and fiction, though many existing robots, mostly in industrial applications, are distant from being intelligent machines (Garcia et al., 2007; Singh et al., 2013).

From a technical perspective, a distinctive definition of a robot has taken time to be globally accepted. In 1988 the International Standard Organization (ISO) states: "An industrial robot is an automatic, freely programmable, multipurpose manipulator, servo-controlled with numerous axes, for the handling of the workpiece, special devices or tools. Variably programmed operations make possible the implementation of a multiplicity of tasks."

Though, different definitions are still utilized. In 1991, the International Federation for the Promotion of Mechanisms and Machine Science previously the International Federation for the Theory of Machines and Mechanisms (IFToMM) provides its definitions. A robot is defined as "Mechanical structure under automatic control that performs actions like handling and locomotion"; and a manipulator is defined as "Device for gripping and effective movements of objects."

Even robotists still utilize their definition for robots to stress some peculiarities. For instance in 2001 from the Institute of Electrical and Electronics Engineers (IEEE) community one can find the statement: "a robot is a machine created as an assemblage of combined links so that they can be expressed into desired positions through a programmable controller and accuracy actuators to execute a variety of tasks." Different terms for robots are still determined from country to country, from application to application and from technical field to technical field (Jacoff et al., 2003; Steels, 2003).

However, a robot or robotic system can be known when it has three key characteristics: re-programming capacity, intelligent capability, and mechanical versatility.

Though, the word 'automaton' can still be appropriately utilized in many cases, when the semantic meaning is entirely understood from the initiating word. The word means "something having the authority of spontaneous motion" and it originates from the Greek word 'automaton,' which is made of the terms 'auto' and 'Matos' that have the meaning of 'self' and 'acting/moving, respectively.

Therefore, the term 'robot' can be given to systems, whose characteristics and meaning are those reported in the aforementioned modern definitions for robots. The automata/automaton term can still be utilized for systems that don't correspond completely to the present robot concept, however, it can still demonstrate autonomous operation ability. Thus, one should appropriately name as robots only those structures that have been designed in the previous five decades. Though those concepts can be known in many systems of the previous and one can still use the word 'robot' in a comprehensive sense when she/he mentions to historical solutions (Wolff and Nordin, 2001; Heinerman et al., 2016).

The idea of an alternate/help for men in heavy/unfriendly work can be found as the beginning of humanity. Thus, machines and tools were conceived, built, and utilized as intermediate solutions with enhancing performance over time. Alternates for manpower were also deliberated in Antiquity. Slaves were the first effective 'intelligent machinery' solutions! However, artificial solutions were also considered, primarily from theoretical viewpoints.

Figure 1.4 displays a Japanese automaton, which is like that of Figure 1.4. The tea maidservant is a design of the 17th century and is skilled in moving and gives a cup of tea. Its design is built on appropriate mechanisms, partially shown in Figure.1.4b), that have been developed in numerous versions until the end of the 17th century. Like the automaton of Figure 1.3, the Japanese automaton of Figure 1.4 was utilized generally for the amusement of the aristocracy. Though it gives a clear instance of the level of proficiency of Japanese culture in the technical area of automata (Watson et al., 2002; Duarte et al., 2016, 2017).

The beginning of the Industrial Revolution progressively promoted the construction and design of automatic devices. Therefore, at the beginning of the 19th century, automatic devices were frequently designed and utilized. Such a great level of knowledge and expertise was attained for the mechanical design of automatic machines in the second half of the 19th century.

Figure 1.3: The writer machines by Jaquet-Droz in 1760: (a) the robot shape; (b) the mechanisms.

Source: https://link.springer.com/chapter/10.1007%2F978-1-4020-2110-7_1.

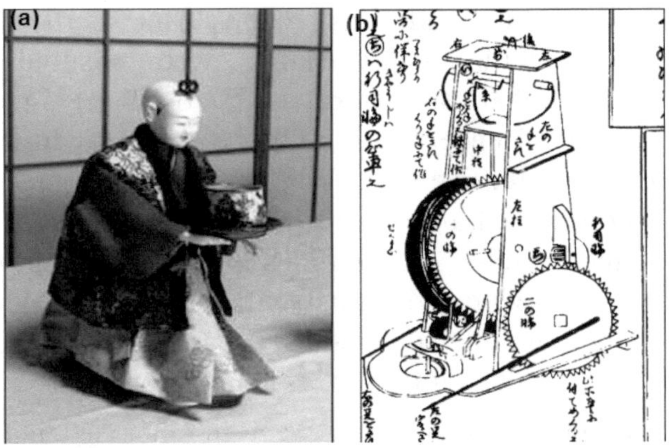

Figure 1.4: The Japanese tea maidservant automaton whose design was created in the 19th century: (a) a modern reconstruction; (b) an original design scheme.

Source: https://link.springer.com/chapter/10.1007%2F978-1-4020-2110-7_1.

Still, in the 20th century, many systems are founded on designs or even creations of mechanisms considered at that time. These automatic machines also demonstrate how a complicated manipulative task can be achieved successfully through mechanisms and not by robots.

A very specific field can be deliberated as the term war machines, as the area of military applications was termed until the 18th century. Meanwhile, early Renaissance the most ingenious designers were employed for construction and conception of latest and more influential weapons, devices for defense and attack activities against fortresses and castles, and latest and more strong defensive civil structures. Numerous new systems were considered, built, designed, and in certain cases even utilized, though several others were re-designed from earlier designs mostly from the rediscovered Roman literature. Now renowned devices are the helicopter, the army tank, and even the first humanoid robot that was developed by Leonardo da Vinci. Also, several other engineers developed weapons and new devices to enhance prevailing military machinery, for instance, the feeding method for automatic guns. However, very often this construction activity and design were not documented since privacy was an important necessity that didn't allow publications or documents on the designed structures, though the benefit of acquired expertise and knowledge also somehow delivered to the non-military uses, as still occurs today (Mondada and Floreano, 1995; Veenstra et al., 2017).

In the start, modern robots were utilized to replace human operators in unhealthy environments and risky actions. But shortly their peculiar characteristics of flexibility and versatility were misused in industrial applications for repeated operations with low-level intellectual and technological contents. Hence, these are the extents of industrial actions in which a 'traditional' extensive utilization of robotized systems and robots had been broadly accepted and is still in the development. Characteristic applications are the handling of hazardous objects, welding, painting, and assembly. In these procedures, the great skill had been developed at such a great level that robots are often deliberated the single solution in those areas (Nolfi and Floreano, 1998; Del Dottore et al., 2018).

Resultantly robots had also been utilized in others extents not only due to acceptance from operators and designers, and improvement required of product quality and productivity but primarily due to the natural progress of automation to robotics as soon as the robots had been developed and considered as industrial machines. Also, substantial has been and is the necessity by the industrial management to demonstrate that production is gained also by utilizing the most advanced and latest equipment for publicity and financial purposes.

Robots are used for technical reasons in non-industrial areas, such as:

- Precise handling;
- Bio-mechanics but industrial applications of robots are experienced in areas such as;
- Loading and unloading of tools and machinery;
- Measure and control of production quality;
- Spatial and submarine explorations;
- Sealing and gluing; and
- Cutting with an extension in all areas of industry.

Substantial robot applications could be deliberated as those for maintenance and servicing or during the installation and assembly of industrial plants. Robots had been utilized in the maintenance and control of nuclear plants for energy generation, and lately even in procedures for deactivated plants. Also, a significant number of robots are utilized in laboratories and research centers working in robotics but also in other areas where they utilize the robots merely as assistant machinery.

The International Federation of Robotics (IFR) lists the number of robots as in Table 1.1 referring to the most industrialized nations by utilizing published data.

Though, it is important to note that users or small companies themselves attain a considerable production of robots. Thus, the data of Table 1.2 for the largest robot manufacturers are listed only to give the size of robot manufacturers and robot production.

Table 1.1: Prevailing Robots in the World (data from IFR publications)

	USA	Italy	UK	France	Japan	Germany	World
Robots in use in 1995	66,000	23,000	8,000	13,000	387,000	51,000	650,000
New robots in 1999	15,063	5,224	1,392	3,092	35,609	10,548	81,508

Great attention and emphasis are previously addressed to inventive solutions for forthcoming applications, mostly at research institutions and universities.

Innovation is predicted not only on the base of new robotic systems but also associated with easy operated and low-cost systems that can be devoted to the friendly utilization of robots.

In specific, regarding short-term development, new robotic systems will be developed for:

- Humanoid robots;
- Robots for medical applications;
- Service robots;
- Robots for agriculture without however mentioning military applications; and
- Submarine and space exploration and work.

Table 1.2: Production of Key Robot Constructors

Robot Constructor	Nationality	New Robots in 1990	In-Service in 1999
ABB	Switzerland-Sweden	3,000	90,000
YASKAWA	Japan	4,400	60,000
COMAU	Italy	500	9,500
NACHI	Japan	1,700	37,000
RENAULT Automation	France	420	7,500
KAWASAKI	Japan	2,000	30,000
HITACHI	Japan	2,600	70,500
KUKA	Germany	800	30,000
CLOOS	Germany	400	8,000
STAUBLIUNIMATION	Switzerland-France	400	8,000
GMF	Japan-USA	3,300	70,000

Service robots would be progressively developed with precise designs but frequently for friendly utilization and co-operation with humans, who might not yet be trained in utilizing robots. This means that excessive efforts should be consumed to make the robot suitable as a friend or company in numerous service applications, through evolving in terms of esthetic appearance and shape. In Japan, they are previously working on authorizing a robot to express some feeling through means of facial expressions and/

or body posture and this they call the area with the multi-meaning term 'Kansei.'

Risky applications, like submarine and space exploration, even work and colonization would need more and more robots, however with a robust design and very innovative features to survive in a tough environment for a long period with very dependable operation ability.

Precise attention had been addressed lately to humanoid robots and great outcomes are predicted to achieve the actuality of the utopia of having a robot that can replace a man in a work action without any authorization. This appears to be a real necessity for the future when low-level works would be performed simply by machinery as humans will not agree to such types of jobs, because they are not capable to provide intellectual satisfaction to educated persons (Kim et al., 2013).

The utilization of robots in the medical area is a typical instance of the latest applications of robots with the abundant potentiality that merely now is being recognized. Many significant achievements are expected generally to help assist surgery, rehabilitation, and other activities with patients. The hard core besides the complicated technical content, comprises of stimulating effective cooperation of two worlds that are antithetic: engineering (robot) and medicine. Indeed, numerous other associations that are not conceivable today would be asked of robotists.

Innovative solutions are predicted mainly for scattering robots more and more by simplifying the implementation and making robotic systems economical. This trend is previously a current problem for robotics but in the upcoming, it could become crucial for the success and even existence of robots.

Therefore, a challenging complication can be deliberated of how to obtain advanced, sophisticated solutions that are needed by technical requirements but with no large operational complexity. Another very significant challenge can be deliberated by the utilization of robots in agriculture. In the 1980s, there were first efforts at robotic harvesting of the horticultural products and fruits from trees, primarily citrus. However, the economic motivation vanished in the 1990s when a great inflow of immigrants made accessible low-cost human operators. Though, in the future it should yet again be of interest, even to encourage robots in agriculture at huge. The utilization of robots in agriculture is a serious aim as it has all the features that still required to be further investigated and specially made practical as they are associated with outdoor applications, however with additional limitations because of the avoidance of loss in cultivation.

Lastly, the most significant challenge for robotists in the upcoming can be considered as to how to assign problems and develop way out in less time in novel fields of applications. New complications and fields will be defined more swiftly, perhaps without the chance that we have currently to study and practice with samples in research centers.

1.4. EXAMPLES AND TECHNICAL CHARACTERISTICS OF INDUSTRIAL ROBOTS

The most significant parameters are summarized in Table 1.3 to provide an outline of the features that can be used for robot selection.

Figures 1.5 is an illustrative instance of a technical advertisement sheet that is generally obtainable for a first choice and assessment of robots since it gives the most significant technical characteristics of the robots.

The models of the control unit and manipulator identify the structure in the robot production. The weights and sizes of the manipulator, power unit, controller, and other sub-systems provide data that are beneficial for proper location and installation of the robot (Megahed, 1993).

The normal application provided by the constructor usually refers to earlier successful experiences, but it doesn't eliminate other applications. The amount of degrees of freedom provides a measure of the kinematic ability of manipulation and usually also indicates the number of independent actuators that work in the manipulator architecture (Chatterjee et al., 2010).

The architecture is categorized by utilizing a classification of its kinematic chain. The manipulator design can be anthropomorphic when it has motion, which is alike to that of the human arm. Amongst those non-anthropomorphic designs, one of the most motivating and most applied is the Selective Compliant Arm for Robot Assembly (SCARA) architecture because of its great performance in object assembling and handling. The workspace is defined as the set of access points by a reference point on the robot extreme. It is of great concern since it gives and explains the manipulation ability of a robot and when it is recognized one can design and appropriately work a work-cell so that the robot can operate accurately the flux of products /objects.

A robot can be programmed as per numerous processes. Usually, the processes used are concerned with the ability of feasible trajectories that can be executed using PTP (point-to-point) or continuous control (Hoffmann et al., 2009).

Table 1.3: Basic Technical Features in Advertisement Sheets for Industrial Robots

Actuators	**Actuator Kind Mechanical Power Operation Range**
Sensors	• Sensors type • Sensing features • Operation range
Kinematic characteristics	• Degrees of freedom • Design Type • Workspace dimensions and shape • Path types • Mobility ranges of extremity and joints.
Control unit	• Hardware type • Programming language • Input/output capability • Input/output features.
General data	• Weight, size, and model of the manipulator • Weight, size, and model of the controller weight, • Size, and model of the power unit • Weight, size, and model, of the other sub-systems
Dynamic characteristics	• Payload (moments and/or mass of inertia) • Maximum acceleration • Maximum velocity • Type of balancing systems
Manipulation	• Accuracy error • Repeatability error

The mobility range is generally provided in terms of the series for the joints. The movement in Cartesian space is specified as a workspace (Roos and Behrens, 1997; Pettersson et al., 2005; Engelberger, 2012).

The payload provides the capability of a robot in transporting weights, but it can also be identified with moments of inertia. The dynamic characteristics are generally stated utilizing extreme values for acceleration and velocity of the robot extremity, which is assessed by taking into account the joint constraints and manipulator dynamics. Utilizing appropriate balancing designs can increase the dynamic and static characteristics. However, the characteristics and type of a balancing design can provide extra limits to the robot operation and appropriate knowledge can assist for optimum utilization of the robot (Seidmann et al., 1984; Bhattacharya et al., 2005; Wallén, 2008).

The kind of actuation (which can be pneumatic, or hydraulic, electric) is given to delivering the required power in the appropriate form and magnitude that are generally specified as general data. The features of the controller are provided in terms of the utilized hardware, interfaces, software performances, and memory capability (Brogårdh, 2007; Karsak, 2008; Arakelian and Briot, 2015).

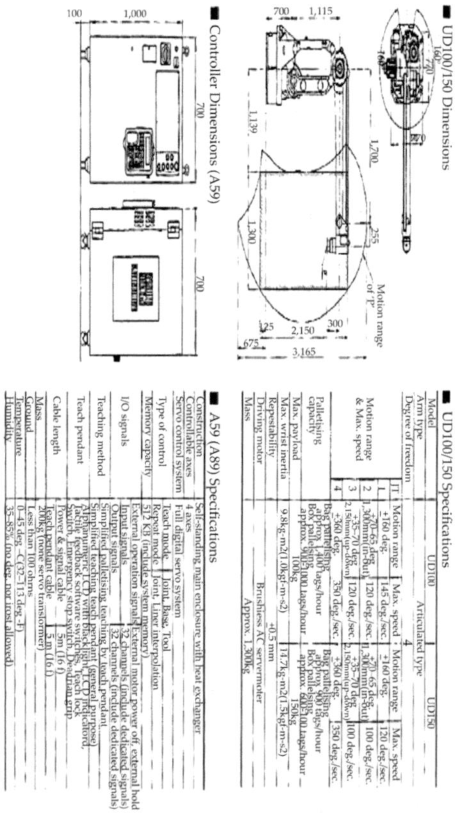

Figure 1.5: Technical advertisement page for a Kawasaki UD100/150 robot and its control unit.

Source: https://link.springer.com/chapter/10.1007%2F978-1-4020-2110-7_1.

The programming language is specified as it can be precise for the robot so that it needs precise training. The input/output proficiency is given with the features of the signals that the regulator can receive and send so that the robot can be incorporated in a robotic work-cell with further machinery under the authorization of a specific control system. The characteristics and types of sensors are beneficial to specify essential control signals and the autonomy level. Usually, force sensors and position sensors are specified to give data on controlled manipulations (Warnecke and Brodbeck, 1977; Ceccarelli, 2003).

Of great significance are the features of the robot manipulation that are usually expressed in terms of repeatability errors and precision. The precision error provides a measure of the uncertainty of extreme positioning for the robot extreme in single programmed operation. The repeatability error provides a measure of the uncertainty of extreme positioning for the robot's extreme repetitive operations. Further, some standards provide reference features and processes for the characterization of robots. International standards are defined and constantly updated by the International Standard Organization (ISO) even referring precisely to robotics (Skibniewski, 1988; Offodile and Madey, 1994).

The selection of a robot can be accomplished by utilizing criteria which primarily concern: technical information of available robots, financial suitability of a robotization, and feasibility of the robotic manipulation.

Certainly, a robot can be nominated not only by utilizing data in the technical advertisement sheet, however even on the base of personal deliberations and mainly earlier experiences with precise robots. These considerations can be a restriction for complete exploiting the potentiality of a robotization and subsequently for an optimum selection of a robot, though they can be of great assistance in attaining the acceptance of a selected robot and robotization (Pires et al., 2007; Thomas et al., 2016).

Thus, it is appropriate to refer to earlier experiences in similar robotizations and productions, but also to enquire about new solutions through designing and operating a robotization with the support of experts in the management and design of the productive procedure. In specific, the design methodology of mimicking human labor with precise robots, though very attractive, sometimes doesn't give the appropriate desired results in terms of product quality and productivity. Indeed, when a robot is utilized one should deliberate the peculiarity of the alternate manual manufacturing, however, effective utilization of robots needs a certain abstraction of the processes to develop even the production for the robots from the commencement (Garcia et al., 2007; Shi et al., 2012).

1.5. EVALUATION OF ROBOTIZATION

Successful utilization of robots in industrial applications rely on numerous aspects from various fields that nonetheless should be appropriately considered both in a panoramic and singular view (Busnel et al., 1999). In this section, only rare considerations are conversed to give an overall view of those relations, which are examined in precise literature to which the reader can refer.

Fundamental relations with the utilization of robots could be known in social impacts, cultural effects, and economic consequences. Regarding financial features, the convenience can be assessed by analyzing the earnings and costs when a robot will replace traditional machinery with human operators. But problems might arise for cost analysis as the market evaluation for labor and primarily for robots is very uncertain (Hacini et al., 2009; Kantelhardt et al., 2013).

Also, in new robot applications, there are inaccessible data from earlier experiences so the cost can merely be considered a desired approximation. Financial suitability can also be considered as intensely associated with quality products and the number of human operators.

It is regularly said that robot utilization causes unemployment, however, this is tough to prove as, though it is true that robots replace human operators with a low level of education, moreover it is considered that automation and generally robotics had encouraged the formation and employment of numerous high-level professionals. Moreover, over-all robot utilization had improved the education level of operators operating with robots (Carrino et al., 2004; Bensignor et al., 2016).

Regarding the production outcomes, it is important to note that quality is not constantly improved but high productivity and primarily uniform and regular production could be accomplished with robot use.

Social effects can be considered as associated with the evolution of the employee's education. In actuality, the utilization of robots needs technicians and primarily engineers with a high class of education so that the ratio of the social classes would alter both in terms of social and numerical components.

It is known that the increase of high-level experts and decrease of human labor alters the composition of the traditional society in production plants with so-termed inverted pyramids, which means that there is no single plant director and numerous operators but numerous directors and rare operators. This social alteration in a production plant modifies the relations and usually, the situation not only in the industrial society however in society at large (Maurin et al., 2006).

Regarding the cultural features, it is obvious that usually, the utilization of robots needs a high level of education. This is accomplished with a longer formation and even with the development of the teaching. On the other hand, the cultural effect can be known also in better education and subsequently the quality of life, as the working environment will be upgraded as a necessity for well-educated people.

The aforementioned some considerations can be further conversed both with negative and positive remarks as in general the utilization of advanced automation and primarily robots don't still give a clear response to the many questions and aspects. This is due to that there is not yet adequate experience, history, and finally tradition. (Koustoumpardis et al., 2007; Belley et al., 2016).

However, it had been considered significant to give a short impression to give a specific panorama of the facts that are involved when utilizing robots. Of course, those facts are discussed in deepness in different precise sources and communities, to which the reader can refer.

However, one of the most significant issues for selecting and utilizing a robot can be deliberated the financial suitability, even from an engineering view (Goffin et al., 2013; Bour et al., 2014).

However, in certain cases, a robotization can be suitable even when the financial evolution provides a negative judgment, asocial considerations or security aspects make it essential. This is the case, for instance of the handling of radioactive hazardous material, or when the risk of breaking-down the procedure could cause environmental harm. In the final case, a robot can be suitably utilized due to sensitivity to the changes in the procedure that can be utilized as an alarm system.

Also, the esthetic reasons or psychological impact for publicity can validate the utilization of robots, irrespective of the direct economic suitability.

REFERENCES

1. Angeles, J., & Angeles, J., (2002). *Fundamentals of Robotic Mechanical Systems* (Vol. 2, pp. 1–33). New York: Springer-Verlag.
2. Arakelian, V., & Briot, S., (2015). *Balancing of Linkages and Robot Manipulators: Advanced Methods with Illustrative Examples* (Vol. 27, pp. 1–33). Springer.
3. Arimoto, S., Kawamura, S., & Miyazaki, F., (1984). Bettering operation of robots by learning. *Journal of Robotic Systems*, *1*(2), 123–140.
4. Balaguer, B., Balakirsky, S., Carpin, S., Lewis, M., & Scrapper, C., (2008). USARSim: A validated simulator for research in robotics and automation. In: *Workshop on Robot Simulators: Available Software, Scientific Applications, and Future Trends at IEEE/RSJ, 1, 1–44.*
5. Bard, J. F., (1986). Evaluating space station applications of automation and robotics. *IEEE Transactions on Engineering Management*, (2), 102–111.
6. Baturone, I., Moreno-Velo, F. J., Sánchez-Solano, S., & Ollero, A., (2004). Automatic design of fuzzy controllers for car-like autonomous robots. *IEEE Transactions on Fuzzy Systems*, *12*(4), 447–465.
7. Belley, A. F., Bouffard, J., Brochu, K., Mercier, C., Roy, J. S., & Bouyer, L., (2016). Development and reliability of a measure evaluating dynamic proprioception during walking with a robotized ankle-foot orthosis, and its relation to dynamic postural control. *Gait and Posture*, *49*, 213–218.
8. Bensignor, T., Morel, G., Reversat, D., Fuks, D., & Gayet, B., (2016). Evaluation of the effect of a laparoscopic robotized needle holder on ergonomics and skills. *Surgical Endoscopy*, *30*(2), 446–454.
9. Bhattacharya, A., Sarkar, B., & Mukherjee, S. K., (2005). Integrating AHP with QFD for robot selection under requirement perspective. *International Journal of Production Research*, *43*(17), 3671–3685.
10. Biggs, G., & MacDonald, B., (2003). A survey of robot programming systems. In: *Proceedings of the Australasian Conference on Robotics and Automation* (Vol. 1, pp. 1–3).
11. Bour, G., Martel, F., Goffin, L., Bayle, B., Gangloff, J., Aprahamian, M., & Egly, J. M., (2014). Design and development of a robotized system Coupled to μCT imaging for intratumoral drug evaluation in a HCC mouse model. *PloS One*, *9*(9), 1–29.
12. Brogårdh, T., (2007). Present and future robot control development—

an industrial perspective. *Annual Reviews in Control*, *31*(1), 69–79.

13. Busnel, M., Cammoun, R., Coulon-Lauture, F., Détriché, J. M., Le Claire, G., & Lesigne, B., (1999). The robotized workstation "MASTER" for users with tetraplegia: Description and evaluation. *Journal of Rehabilitation Research and Development*, *36*(3), 217–229.
14. Carrino, L., Polini, W., & Sorrentino, L., (2004). Method to evaluate winding trajectories in robotized filament winding. *Journal of Composite Materials*, *38*(1), 41–56.
15. Ceccarelli, M., (2003). Robotic teachers' assistants. *IEEE Robotics and Automation Magazine*, *10*(3), 37–45.
16. Chapman, T., (2003). Lab automation and robotics: Automation on the move. *Nature*, *421*(6923), 661–663.
17. Chatterjee, P., Athawale, V. M., & Chakraborty, S., (2010). Selection of industrial robots using compromise ranking and outranking methods. *Robotics and Computer-Integrated Manufacturing*, *26*(5), 483–489.
18. Chui, M., Manyika, J., & Miremadi, M., (2015). Four fundamentals of workplace automation. *McKinsey Quarterly*, *29*(3), 1–9.
19. Cusack, M., (1994). Automation and robotics the interdependence of design and construction systems. *Industrial Robot: An International Journal*, *21*(4), 10–14.
20. Daerden, F., Lefeber, D., Verrelst, B., & Van, H. R., (2001). Pleated pneumatic artificial muscles: Actuators for automation and robotics. In: *2001 IEEE/ASME International Conference on Advanced Intelligent Mechatronics. Proceedings (Cat. No. 01TH8556)* (Vol. 2, pp. 738–743). IEEE.
21. Del, D. E., Sadeghi, A., Mondini, A., Mattoli, V., & Mazzolai, B., (2018). Toward growing robots: A historical evolution from cellular to plant-inspired robotics. *Frontiers in Robotics and AI*, *5*, 16.
22. Dietrich, D., Bruckner, D., Zucker, G., Muller, B., & Tmej, A., (2009). Psychoanalytical model for automation and robotics. In: *AFRICON 2009* (Vol. 1, pp. 1–8). IEEE.
23. Duarte, M., Costa, V., Gomes, J., Rodrigues, T., Silva, F., Oliveira, S. M., & Christensen, A. L., (2016). Evolution of collective behaviors for a real swarm of aquatic surface robots. *PloS One*, *11*(3), 1–40.
24. Duarte, M., Gomes, J., Oliveira, S. M., & Christensen, A. L., (2017). Evolution of repertoire-based control for robots with complex locomotor systems. *IEEE Transactions on Evolutionary Computation*,

22(2), 314–328.

25. Elattar, S. M. S., (2008). Automation and robotics in construction: Opportunities and challenges. *Emirates Journal for Engineering Research*, *13*(2), 21–26.
26. Elkmann, N., Felsch, T., Sack, M., Saenz, J., & Hortig, J., (2002). Innovative service robot systems for facade cleaning of difficult-to-access areas. In: *IEEE/RSJ International Conference on Intelligent Robots and Systems* (Vol. 1, pp. 756–762). IEEE.
27. Engelberger, J. F., (2012). *Robotics in Practice: Management and Applications of Industrial Robots* (Vol. 1, pp. 1–33) Springer Science and Business Media.
28. Everett, J. G., & Slocum, A. H., (1994). Automation and robotics opportunities: Construction versus manufacturing. *Journal of Construction Engineering and Management*, *120*(2), 443–452.
29. Garcia, E., Jimenez, M. A., De Santos, P. G., & Armada, M., (2007). The evolution of robotics research. *IEEE Robotics and Automation Magazine*, *14*(1), 90–103.
30. Goffin, L., Bour, G., Martel, F., Nicolau, S., Gangloff, J., Egly, J. M., & Bayle, B., (2013). Design and in vivo evaluation of a robotized needle insertion system for small animals. *IEEE Transactions on Biomedical Engineering*, *60*(8), 2193–2204.
31. Grift, T., Zhang, Q., Kondo, N., & Ting, K. C., (2008). A review of automation and robotics for the bio-industry. *Journal of Biomechatronics Engineering*, *1*(1), 37–54.
32. Hacini, L., Van, L. N., & Bocher, P., (2009). Evaluation of residual stresses induced by robotized hammer peening by the contour method. *Experimental Mechanics*, *49*(6), 775.
33. Hasegawa, Y., (2006). Construction automation and robotics in the 21st century. In: *International Symposium on Automation and Robotics in Construction (ISARC)* (Vol. 1, pp. 565–568).
34. Heinerman, J., Zonta, A., Haasdijk, E., & Eiben, A. E., (2016). On-line evolution of foraging behavior in a population of real robots. In: *European Conference on the Applications of Evolutionary Computation* (Vol. 1, pp. 198–212). Springer, Cham.
35. Heinzmann, J., & Zelinsky, A., (1999). The safe control of human-friendly robots. In: *Proceedings 1999 IEEE/RSJ International Conference on Intelligent Robots and Systems. Human and Environment*

Friendly Robots with High Intelligence and Emotional Quotients (Cat. No. 99CH36289) (Vol. 2, pp. 1020–1025). IEEE.

36. Hoffmann, A., Angerer, A., Ortmeier, F., Vistein, M., & Reif, W., (2009). Hiding real-time: A new approach for the software development of industrial robots. In: *2009 IEEE/RSJ International Conference on Intelligent Robots and Systems* (Vol. 1, pp. 2108–2113). IEEE.
37. Hornby, G. S., Lipson, H., & Pollack, J. B., (2001). Evolution of generative design systems for modular physical robots. In: *Proceedings 2001 ICRA. IEEE International Conference on Robotics and Automation (Cat. No. 01CH37164)* (Vol. 4, pp. 4146–4151). IEEE.
38. Jacoff, A., Weiss, B., & Messina, E., (2003). *Evolution of a Performance Metric for Urban Search and Rescue Robots* (Vol. 1, pp. 1–33). National Institute of standards and technology Gaithersburg Md.
39. Kamimura, A., Kurokawa, H., Toshida, E., Tomita, K., Murata, S., & Kokaji, S., (2003). Automatic locomotion pattern generation for modular robots. In: *2003 IEEE International Conference on Robotics and Automation (Cat. No. 03CH37422)* (Vol. 1, pp. 714–720). IEEE.
40. Kamimura, A., Kurokawa, H., Yoshida, E., Murata, S., Tomita, K., & Kokaji, S., (2005). Automatic locomotion design and experiments for a modular robotic system. *IEEE/ASME Transactions on Mechatronics, 10*(3), 314–325.
41. Kantelhardt, S. R., Finke, M., Schweikard, A., & Giese, A., (2013). Evaluation of a completely robotized neurosurgical operating microscope. *Neurosurgery, 72*(suppl_1), A19–A26.
42. Karsak, E. E., (2008). Robot selection using an integrated approach based on quality function deployment and fuzzy regression. *International Journal of Production Research, 46*(3), 723–738.
43. Karwowski, W., Rahimi, M., & Mihaly, T., (1988). Effects of computerized automation and robotics on safety performance of a manufacturing plant. *Journal of Occupational Accidents, 10*(3), 217–233.
44. Keramas, J. G., Schin, T., McAvey, F., & Main, L., (1998). *Robot Technology Fundamentals* (Vol. 1, pp. 1–33). Delmar Learning.
45. Kim, S., Laschi, C., & Trimmer, B., (2013). Soft robotics: A bioinspired evolution in robotics. *Trends in Biotechnology, 31*(5), 287–294.
46. Kolluru, R., Valavanis, K. P., Smith, S. S., & Tsourveloudis, N., (2000). Design fundamentals of a reconfigurable robotic gripper system. *IEEE*

Transactions on Systems, Man, and Cybernetics-Part A: Systems and Humans, *30*(2), 181–187.

47. Koustoumpardis, P. N., Fourkiotis, J. S., & Aspragathos, N. A., (2007). Intelligent evaluation of fabrics' extensibility from robotized tensile test. *International Journal of Clothing Science and Technology, 1*, 1–33.
48. Koza, J. R., & Rice, J. P., (1992). Automatic programming of robots using genetic programming. In: *AAAI* (Vol. 92, pp. 194–207).
49. Li, Z., & Ge, S. S., (2013). *Fundamentals in Modeling and Control of Mobile Manipulators* (Vol. 49, pp. 1–33). CRC Press.
50. Maurin, B., Bayle, B., Gangloff, J., Zanne, P., De Mathelin, M., & Piccin, O., (2006). A robotized positioning platform guided by computed tomography: Practical issues and evaluation. In: *Proceedings 2006 IEEE International Conference on Robotics and Automation, 2006. ICRA 2006.* (Vol. 1, pp. 251–256). IEEE.
51. Megahed, S. M., & Megahed, S. M., (1993). *Principles of Robot Modeling and Simulation* (Vol. 1, pp. 1–33). J. Wiley.
52. Meystel, A., (1991). *Autonomous Mobile Robots: Vehicles with Cognitive Control* (Vol. 1, pp. 1–33). World scientific.
53. Mondada, F., & Floreano, D., (1995). Evolution of neural control structures: Some experiments on mobile robots. *Robotics and Autonomous Systems*, *16*(2–4), 183–195.
54. Montemerlo, M., (2001). 25 years of space automation and robotics at NASA: An historical perspective. In: *Proceeding of the 6th International Symposium on Artificial Intelligence and Robotics and Automation in Space: I-SAIRAS, 1,* 1–33.
55. Mooring, B. W., Roth, Z. S., & Driels, M. R., (1991). *Fundamentals of Manipulator Calibration* (Vol. 1, p. 329). New York: Wiley.
56. Nolfi, S., & Floreano, D., (1998). Coevolving predator and prey robots: Do "arms races" arise in artificial evolution? *Artificial Life*, *4*(4), 311–335.
57. Offodile, F. O., & Madey, G. R., (1994). A CLIPS-based expert system for the evaluation and selection of robots. In: *Third CLIPS Conference Proceedings* (Vol. 1, p. 131). NASA Center for AeroSpace Information.
58. Paulson, Jr. B. C., (1985). Automation and robotics for construction. *Journal of Construction Engineering and Management*, *111*(3), 190–207.

59. Pettersson, M., Krus, P., & Andersson, J., (2005). On optimal drive train design in industrial robots. In: *2005 IEEE International Conference on Industrial Technology* (Vol. 1, pp. 254–259). IEEE.

60. Pires, J. N., Godinho, T., Nilsson, K., Haage, M., & Meyer, C., (2007). Programming industrial robots using advanced input-output devices: Test-case example using a CAD package and a digital pen based on the anoto technology. *International Journal of Online Engineering, 3*(3), 7.

61. Poole, H. H., (2012). *Fundamentals of Robotics Engineering* (Vol. 1, pp. 1–30). Springer Science and Business Media.

62. Richard, R. B., (2005). Industrialized building systems: Reproduction before automation and robotics. *Automation in Construction, 14*(4), 442–451.

63. Roos, E., & Behrens, A., (1997). Off-line programming of industrial robots—Adaptation of simulated user programs to the real environment. *Computers in Industry, 33*(1), 139–150.

64. Saunders, K. C., (2004). Automation and robotics in ADME screening. *Drug Discovery Today: Technologies, 1*(4), 373–380.

65. Seidmann, A., Arbel, A., & Shapira, R., (1984). A two-phase analytic approach to robotic system design. *Robotics and Computer-Integrated Manufacturing, 1*(2), 181–190.

66. Shi, J., Jimmerson, G., Pearson, T., & Menassa, R., (2012). Levels of human and robot collaboration for automotive manufacturing. In: *Proceedings of the Workshop on Performance Metrics for Intelligent Systems* (Vol. 1, pp. 95–100).

67. Singh, B., Sellappan, N., & Kumaradhas, P., (2013). Evolution of industrial robots and their applications. *International Journal of Emerging Technology and Advanced Engineering, 3*(5), 763–768.

68. Skibniewski, M. J., (1988). Framework for decision-making on implementing robotics in construction. *Journal of Computing in Civil Engineering, 2*(2), 188–201.

69. Skibniewski, M., & Hendrickson, C., (1990). Automation and robotics for road construction and maintenance. *Journal of Transportation Engineering, 116*(3), 261–271.

70. Slaughter, E. S., (1997). Characteristics of existing construction automation and robotics technologies. *Automation in Construction, 6*(2), 109–120.

71. Smithers, T., (1997). Autonomy in robots and other agents. *Brain and Cognition*, *34*(1), 88–106.
72. Song, G., Wang, H., Zhang, J., & Meng, T., (2011). Automatic docking system for recharging home surveillance robots. *IEEE Transactions on Consumer Electronics*, *57*(2), 428–435.
73. Steels, L., (1995). When are robots intelligent autonomous agents? *Robotics and Autonomous Systems*, *15*(1, 2), 3–9.
74. Steels, L., (2003). Evolving grounded communication for robots. *Trends in Cognitive Sciences*, *7*(7), 308–312.
75. Suh, S. H., Woo, I. K., & Noh, S. K., (1991). Automatic trajectory planning system (ATPS) for spray painting robots. *Journal of Manufacturing Systems*, *10*(5), 396–406.
76. Thomas, C., Matthias, B., & Kuhlenkötter, B., (2016). Human-robot-collaboration-new applications in industrial robotics. In: *International Conference on Competitive Manufacturing* (Vol. 1, pp. 293–299).
77. Thomas, U., & Wahl, F. M., (2001). A system for automatic planning, evaluation and execution of assembly sequences for industrial robots. In: *Proceedings 2001 IEEE/RSJ International Conference on Intelligent Robots and Systems. Expanding the Societal Role of Robotics in the Next Millennium (Cat. No. 01CH37180)* (Vol. 3, pp. 1458–1464). IEEE.
78. Todd, D. J., (2012). *Fundamentals of Robot Technology: An Introduction to Industrial Robots, Tele-operators and Robot Vehicles* (Vol. 1, pp. 1–33). Springer Science & Business Media.
79. Veenstra, F., Faina, A., Risi, S., & Stoy, K., (2017). Evolution and morphogenesis of simulated modular robots: A comparison between a direct and generative encoding. In: *European Conference on the Applications of Evolutionary Computation* (Vol. 1, pp. 870–885). Springer, Cham.
80. Wallén, J., (2008). *The History of the Industrial Robot* (Vol. 1, pp. 1–22*)*. Linköping University Electronic Press.
81. Warnecke, H. J., & Brodbeck, B., (1977). Analysis of industrial robots on a test stand. *Industrial Robot: An International Journal, 1*, 1–33.
82. Watson, R. A., Ficici, S. G., & Pollack, J. B., (2002). Embodied evolution: Distributing an evolutionary algorithm in a population of robots. *Robotics and Autonomous Systems*, *39*(1), 1–18.

83. Wolff, K., & Nordin, P., (2001). Evolution of efficient gait with humanoids using visual feedback. In: *Proceedings of the 2nd IEEE-RAS International Conference on Humanoid Robots, Humanoids* (Vol. 1, pp. 99–106).

84. Wu, N., & Zhou, M., (2007). Deadlock resolution in automated manufacturing systems with robots. *IEEE Transactions on Automation Science and Engineering*, *4*(3), 474–480.

85. Xie, M., (2003). *Fundamentals of Robotics: Linking Perception to Action* (Vol. 54, pp. 1–44). World Scientific.

86. Zavadskas, E. K., (2010). Automation and robotics in construction: International research and achievements. *Automation in Construction*, *19*(3), 286–290.

87. Zhang, Y., Roufas, K., Eldershaw, C., Yim, M., & Duff, D., (2003). Sensor computations in modular self-reconfigurable robots. In: *Experimental Robotics VIII* (Vol. 8, pp. 276–286). Springer, Berlin, Heidelberg.

Chapter 2

Introduction to Agricultural Automation

CONTENTS

2.1. INTRODUCTION

Agricultural productivity had considerably increased through the years by mechanization, automation, and intensification. This comprises automated farming apparatus for animal systems, field operations, and developing systems (irrigation systems, greenhouse climate control). Beginning of automation into agriculture had reduced production costs, decrease the drudgery of physical labor, enhanced quality of fresh yield, and upgraded environmental control (Auernhammer et al., 1991; O'Connor et al., 1996; Stombaugh et al., 1998). Disparate industrial applications, which deal with repetitive, well-defined, simple, and a significance known tasks, automation in agriculture need advanced technologies to deal with the highly flexible environment and complex and produce. Agricultural products are natural objects which had a high level of variability as a consequence of genetic and environmental variables (Bell, 2000; Wilson, 2000; Isidori and De Persis, 2009). The agricultural environment is loosely structured and complicated with large differences amongst fields and even inside the same field. Fundamental technologies need to be developed to resolve difficult problems like variability in products, delicate products, continuously changing conditions and environment (location, soil properties, shape, weather, and size) and hostile environmental situations (extreme temperature, dirt dust, and humidity). Intelligent control structures are essential for real-time, dynamic interpretation of the objects and the environment (Keicher and Seufert, 2000; Torii, 2000; Noguchi et al., 2002). When matched with industrial automation structures, precision necessities in agricultural automation structures might be much lower. As the product being allocated with is of comparatively low cost, the cost of the automated structure needs to below for it to be economically accepted. The seasonal nature of agriculture makes it tough to attain the high utilization placed in manufacturing industries (Ferguson, 1925; Schueller, 1992; Reid et al., 2000).

2.2. FIELD MACHINERY

The utilization of machinery in agriculture had a long history, but the most substantial developments happened during the twentieth century with the beginning of tractors. As early as 1903, the initial farm tractor motorized through an internal combustion engine was manufactured by Hart-Parr Company. Utilizing its assembly line methods, Henry Ford and Son Corporation initiated large production of Fordson tractors in 1917 (Reid et al., 1985; Gordon and Holmes, 1988; Noguchi et al., 1997). The commercial achievement of tractors sparked other inventions as well. In 1924, the International Harvester Company presented a power takeoff device that permitted power from a tractor engine to be transferred to the attached device like a mechanical reaper. Deere and Company surveyed in 1927 with power lift equipment that lowered and raised hitched implements at the close of each row. Rubber wheels were first developed and utilized for tractors in 1932 to increase fuel economy and traction (Gerrish et al., 1985; Marchant and Brivot, 1995; Marchant, 1996). Powered and pulled by tractors, a progressively extensive variety of farm implements were developed in the twentieth century to systematize crop production in every phase, from tillage, planting, to harvesting. Harvesting equipment lagged only tractors insignificance. Initial harvesters for small-grain crops were powered by tractors and pulled by tractors PTO (power takeoff). The growth of a self-propelled associate in 1938 by Massey Harris marked substantial progress in enhancing productivity. The self-propelled associate incorporated numerous functions like grain gathering, vehicle propulsion, and grain threshing into an all-in-one component for improved operation effectiveness (Billingsley, and Schoenfisch, 1997; Farrell et al., 2000; Han et al., 2004). The automation of harvesting other crops comprised the advancement of mechanical spindle cotton pickers in 1943 and mechanical hay balers in the 1930s. Combines, tractors, and other farm equipment were constantly refined throughout the second half of the twentieth century to be more productive, user-friendly, and efficient. The achievement of agricultural mechanization had created a strong basis for automation (Figure 2.1).

Figure 2.1: Agricultural wheat harvester.

Source: https://www.iso.org/news/2015/12/Ref2033.html.

Automation enhances the productivity of agricultural machinery by enhancing efficiency, precision, and reliability and decreasing the requirement of human intervention (Guo et al., 2002). This is accomplished by adding controls and sensors. The combination of sensors with mechanical actuation could be established in numerous agricultural operations like vision-guided tractors, product grading systems, planters, and harvesters, automating growing conditions, irrigation, and fertilizer applicators (Figure 2.2).

Figure 2.2: Cotton picking machine-agricultural automation.

Source: https://openoregon.pressbooks.pub/envirobiology/chapter/9-3-conventional-agriculture/.

The past of automation for agricultural machinery is nearly as old as agricultural modernization. Two ingenious instances in the early twentieth century were the implement draft control structure by Ferguson in 1925 and the self-leveling structure for hillside combines by Holt Co. in 1891. Early automation structures mainly utilized hydro-mechanical and mechanical control devices. Since the 1960s, electronics growth for control and monitoring had ruled machine designs and has led to enhanced machinery intelligence and automation. Mechatronics technology, a combination of mechanics, computing, and electronics is often used to design of latest automation systems. Automation in modern agricultural machines is much more complex than a single control action; for instance, the latest combine harvester had automatic control of travel speed, rotor speed, concave opening, header height reel speed, and sieve opening to improve the whole harvest process. Farm machinery comprises transport vehicles and tractors, fertilizer applicators, tillage, and seeding machines, and plant safety application equipment, a device for post-harvest preservation, harvesters, a, and treatment of yield. Automation and mechanization instances can be found in numerous of these machines (Levin et al., 2010). Though, the extensive variety of agricultural structures and their diversity all over the world makes it hard to simplify the application of control and automation. Thus, only a single type of automation (automated navigation) of agricultural vehicles would be demonstrated here. Automated vehicle navigation structures comprise the automatic steering system, operator-assisted steering system, and autonomous structure. These systems could release the vehicle operator of the monotonous and repetitive steering operation. Automatic guidance had been the most vigorous research area in the automation past of agricultural machinery. With the beginning of the global positioning system (GPS) to agriculture at the end of the 1980s, automatic guidance technology had been effectively commercialized. Nowadays, auto guidance is the fastest enhancing section in the agricultural machinery industry. The following segments debate the principles of auto-guidance systems, the access technologies, and instance of precise auto-guidance systems.

2.2.1. Automatic Guidance of Agricultural Vehicles

For numerous agricultural operations, an operator is needed to perform two basic tasks simultaneously: operating the equipment and steering the vehicle. The need to release the operator of constantly making steering changes had been the main motive for the development of automatic guidance systems. Outstanding references to automatic vehicle management research in Japan, Europe, Canada, and the USA can be found in Keicher and Seufert (2000), Torii (2000), Reid et al. (2000), and Wilson (2000). Figure 2.3 shows a distinctive auto-guidance system that comprises a steering angle sensor, a position sensor, a path planner, and a steering actuator as the hardware components, a steering controller as the software components, and a navigation controller. The path planner provides the planned (or desired) vehicle position. This desired position is matched with the measured position provided by the position sensor. The navigation controller estimates the desired steering control angle founded on the difference in the measured and desired positions. Lastly, the steering controller utilizes the difference in the measured and desired steering angles to estimate an executing steering control signal and refers it to drive the steering actuator. The latest agricultural vehicles frequently employ E/H (electrohydraulic) steering systems. Improvements in each of the system parts are described in detail below (Grewal et al., 2020).

2.2.1.1. Position Sensing

The position sensing structure measures vehicle location comparative to a reference frame and gives inputs to the navigation controller. Maximum agricultural guidance applications need position measurement in 2-D (two dimensional) space. Also, heading, rotational movements (pitch, yaw, roll) vehicle speed are often required by the navigation controller.

Guidance exactness is the main factor in picking a position sensor. Auernhammer et al. (1991) recommended three stages of accuracy needed for different farming operations: 1 meter for rough operations (weed scouting, soil sampling), 10 centimeters for fine operations (soil cultivation, pesticide application), and 1 centimeter for precise operations (plowing, planting). Different position sensors are nominated in a guidance system to fulfill the precision requirements for the different farming operations. Overall, there are three types of positioning methods: absolute positioning, sensor fusion, and relative positioning.

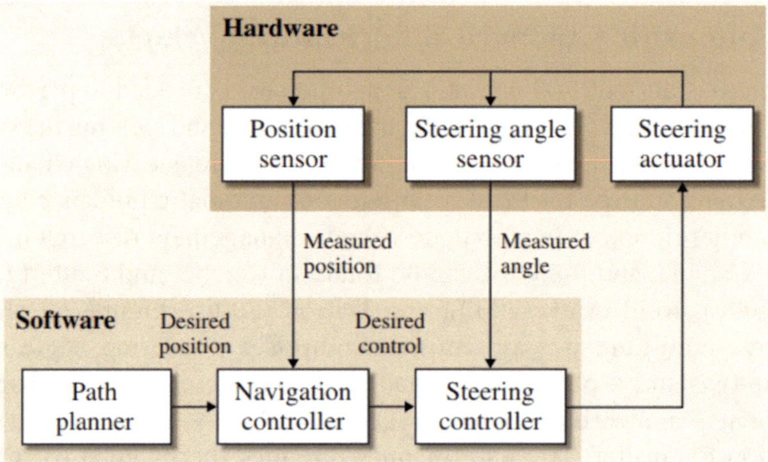

Figure 2.3: Mechanism of a typical auto-guidance system.

Source: https://www.semanticscholar.org/paper/Automation-i-63.-Automation-in-Agriculture-Edan-Han/774ae073f1ba426b147b4c0228e40b31b7cf08f0.

2.2.1.2. Path Planning

Path planning is the generation of two-dimensional sequenced places or trajectories for the automatic vehicle. The sequenced positions interpret for the vehicle kinematics like the lowest turn radius and other constraints. Maximum agricultural operations, like tillage, planting, spraying, and harvesting, needs the vehicle to travel the whole field with parallel paths at a stable spacing equaling the implement thickness. Planning such paths is termed coverage path planning. Coverage path planning includes two steps. Step 1 is to mold a field into sub-regions. An optimum travel direction is placed for every sub-region. Step 2 is to find the optimum coverage pattern inside every sub-region. Numerous different algorithms have been made for coverage path planning (Choset, 2001; Cohen and Edan, 2008).

Trapezoidal decomposition is a widespread method for subdividing the field. The trapezoids are then combined into larger blocks and the choice is made utilizing certain standards which take into consideration the route length and the area of the block and the effectiveness of driving (Oksanen et al., 2005); Jin and Tang (2006) utilized a geometric model to demonstrate the full coverage path planning issues. The algorithm was proficient in finding a worldwide optimal decomposition for a specified field and the direction of the boustrophedon paths for every sub-region. The search process of the algorithm is directed through a customized cost function that merges different cost criteria, and a divide-and-conquer strategy is implemented (Shani, 1980; Wu et al., 1999; Zhang, 1999).

A graphical method is often utilized to find the enhanced coverage pattern inside a sub-region (Huang et al., 1986; Cao et al., 1988). The processes comprise dividing the area, building a divider graph, and examining the partition graph. Heuristic functions are utilized in the searching procedure to prune the search tree promptly so the optimized resolution could be found within a rational time. In the instance of many vehicles working in the similar region, Gray (2002) produced a path planning method (Figure 2.4).

Figure 2.4: A John Deere 8000 series tractor fitted with the green star Auto-Trac assisted steering structure.

Source: https://www.researchgate.net/publication/226815941_Automation_in_Agriculture.

2.2.2. Autonomous Agricultural Vehicles and Robotic Field Operations

A self-governing vehicle must be capable to function without an operator. In addition to steering, it must execute other tasks that a human operator usually does: detecting and avoiding unidentified objects, functioning at a safe speed, and executing implement tasks while driving. Developing human intelligence for the self-governing vehicle is a challenging work (Park and Nikravesh, 1996; Zhang and Qiu, 2004).

Self-governing vehicles working in an amorphous agricultural environment must utilize control systems and sophisticated sensing to be capable to react to any unexpected events. A usual unexpected event is the occurrence of an animal or human in front of the vehicle (Qiu et al., 2001). Development of vehicle protection systems is significant to the positioning of the autonomous vehicles. Several technologies had been inspected for providing vehicle protection. Guo et al. (2002) utilized two ultrasonic sensors to identify a human being. The reliable detection extent was up to

4.6 meters for moving objects and 7.5 meters for static objects under field circumstances. Oksanen et al. (2005) utilized a binocular stereo camera to identify a person standing in front of a vehicle. The system was capable to find the person's comparative motion status (heading and speed) comparative to the vehicle at a distance extent from 3.4–13.4 m. Kise et al. (2005) utilized a laser range locater to approximate the relative motion of a tractor hurdle. Usually, ultrasonic sensors are less cost however, their detection range is little. Stereo cameras are not trustworthy under altering lighting conditions. Presently, the most trusted technology is the laser rangefinder, however, its utilization is restricted to research vehicle platforms because of high costs. Multiple stages of structure redundancy must be deliberated into the vehicle, which often needs multiple protection sensors. Progress of the robotic field task is an essential part of self-governing vehicles. To utilize an autonomous vehicle, chores must be automated also. Over the years, agricultural machines had evolved to put up the automated control of chores. Microprocessor-based electronics control is substituting mechanical control, an electro-hydraulically powered actuators are chosen over mechanically powered actuators. The implementation of CAN bus standards (ISO 11783, DIN 9684, SAE J1939) in the agricultural machine industry had permitted networking of multiple control structures.

Task automation instances could be found in numerous modern agricultural equipment. Examples comprise map-based automatic spraying of chemicals and fertilizer on sprayers, and headland management systems (HMS) for automated sequencing of tractor functions usually related to headland turns. Matsuo et al. (2002) explained a tilling robot that was capable to do tillage, seedling, and soil paddling processes.

Reid (2004) conversed some challenges associated with the development of intelligent agricultural equipment and machinery. Currently, robotic field operations and autonomous agricultural vehicles are still not durable and reliable enough to meet the needs of the agricultural industry and its clients.

2.2.2.1. Robotic Harvester

A robotic harvester, termed Demeter (Figure 2.5), had been created by the Carnegie Mellon University Robotics Institute for automatic harvesting of windrowed crops. The robot displayed place was a New Holland 2550 self-propelled windrower fitted with two color cameras, an inertial navigation system (INS), and DGPS.

Figure 2.5: The robotic harvester (Demeter).

Source: http://www.technovelgy.com/ct/Science-Fiction-News.asp?NewsNum=177.

The camera system identified the uncut/cut edge of the crop, which provided a comparative directrix for the harvester to follow. The camera system was similarly utilized to detect possible obstacles for vehicle protection. GPS data was attached with vision data for direction. In addition to steering, the header height and speed of the harvester were also automatically organized. In 1997, the Demeter freely harvested 100 acres of alfalfa in a constant run (eliminating stops for refueling). During 1998, the Demeter harvested an additional 120 acres of the crop, cutting in both alfalfa and Sudan fields (Ollis and Stentz, 1997; Pilarski et al., 2002).

2.2.2.2. Autonomous Tractor

An autonomous tractor had been mutually developed by Autonomous Solutions Inc. and John Deere for automated mowing, tillage, and spraying in orchards (Figure 2.6). The robot displayed place was a John Deere 5000 series tractor with substantial amendments. The system components comprised a base station, a vehicle, and a mobile control unit; all were interconnected through a wireless CAN system. An INS and DGPS were utilized as positioning sensors. Vehicle controls comprised PTO, clutch, brake, steering, throttle, and three-point hitch. A long-range hurdle detection system was suggested for vehicle protection. One of the significant developments in the project was the mission and path planning, which comprised dynamic re-planning for dynamic service occasions. The structure design followed an industry joint architecture for unmanned ground systems (JAUGS) architecture. A proof-of-concept structure was developed and effectively demonstrated, however, the production decision was not done, mainly because of safety concerns.

Figure 2.6: A John Deere 5000N independent orchard tractor (courtesy of Deere and company).

Source: https://static.agcanada.com/wp-content/uploads/2017/07/GNN170718.pdf.

2.2.2.3. Small Robotic Platforms

In agriculture, small robots could be utilized for numerous field tasks like a collection of plant or soil samples and detection of insect, weed, or plant stress. When equipped with an appropriate actuator and larger energy source, they could also be utilized for localized treatments like mechanical in row weeding or spot-spraying of chemicals. Several little robots had been developed, primarily at research institutes and universities (Mast et al., 2012). Astrand and Baerveldt (2002) developed an independent robot for mechanical weed control in outside surroundings. The robot engages a grey-level vision structure to direct itself laterally to the crop rows and a second, color-based vision structure to detect the weed and to control a weeding device that eliminates the weed inside the row of crops. A plant nurturing robot, HortiBot, was developed in Denmark as a device carrier for accurate weeding (Jorgensen et al., 2006). The HortiBot is a radio-controlled incline mower (Spider ILD01, Dvorák Machine Division, Czech Republic) fitted with a robotic accessory set. A commercial stereo vision structure was executed for automatic guidance inside plant rows.

2.2.3. Future Directions and Prospects

Farm productivity had enhanced considerably throughout the last century. Currently, less than 3% of the United States population works in agriculture, however, they produce more than sufficient food for the whole nation. Agricultural automation had played an important role in attaining this miracle. As the next phases to automation, mechanization, and robotization of farm processes can consequence in additional productivity up-gradation.

Auto guidance will remain to be the main emphasis of forthcoming e development. The agricultural industry is currently developing new structures for automation beyond vehicle auto-steering. Headland management and Implement guidance are two instances. An implement guidance structure automatically steers together tractor and implement and keeps the implement on the chosen path. This assists to overcome implement drift on contour field or hillsides situations. The headland management structure automatizes implement controls (e.g., to lower or raise the implement) and makes automatic turns at the interior field and headland boundaries. Additional guidance technologies that are near to commercialization comprise path or mission planning that produces the most effective coverage paths for multiple or single vehicles, sensor fusion that engages a multitude of corresponding positioning sensors to expand system reliability, and leader-follower structures for multiple vehicle navigation and control, as in the situation of combine harvester operation. Precision farming had developed an area of huge growth and excitement since the 1980 s. The main conception in precision farming is to achieve crop production at the subfield level. The labor-intensive nature of precision farming applies brings a huge requirement for automated equipment and machines. Variable-rate application and yield mapping systems are nowadays commercially available. In the upcoming, independent field, scout vehicles are required for crop scouting, soil sampling, and real-time facts collection. Small robots are chosen for individual plant care like selective crop harvesting and precision weed control. Since precision farming is deliberated as the future of agriculture, robotics, and automation technologies will become an enormous part of agriculture production in the twenty-one century (Sorensen et al., 2006) (Figures 2.7 and 2.8).

Figure 2.7: Sprinkler irrigation.

Source: https://www.sciencedirect.com/topics/agricultural-and-biological-sciences/sprinkler-irrigation.

Figure 2.8: Center pivot by drop sprinklers.

Source: http://nelsonirrigation.com/products/application/center-pivot.

2.3. IRRIGATION SYSTEMS

Irrigation is the additional application of water to the soil for supporting in growing crops. It is utilized mainly to substitute missing rainfall for the field crops, and to provide water to crops growing in secure environments like greenhouses. The main aim is to supply the needed amount of water to the plants at the exact time (Norremark et al., 2006). The types of irrigation methods vary in how the water is dispersed inside the field. In surface irrigation structures, water moves on the land by gravity and penetrates the soil. Surface irrigation systems comprise border-strip, furrow, basin irrigation, and Wildlife Service. Localized irrigation systems dispense water in piped networks through pressure, and the water is used locally to the plant and in the field. Localized systems comprise drip, bubble, spray, and sprinkler systems. Automation gives effective on-farm utilization of labor and water for all techniques by enabling rate, duration of water supply and flexible frequency with control of the irrigator at the exact application point (Merriam et al., 2007).

2.3.1. Types of Irrigation Systems

Flood control automation comprises finest gate operation of the irrigation reservoirs (S. Kim and P. Kim, 2004); surge flooding, which allows the discharge of water at planned intervals; canal water level and telemetering of paddy ponding deepness (Briskin et al., 2020), which could be utilized to capture surplus and pump it back again into the field for reutilization; and accuracy control of inflow rate utilizing ground-based remote-sensing response control structures (Lam et al., 2006, 2007). The position of the progress of water laterally to the furrow could be determined through contact-type sensors physically positioned in the furrow and recently through imaging systems (Figure 2.9).

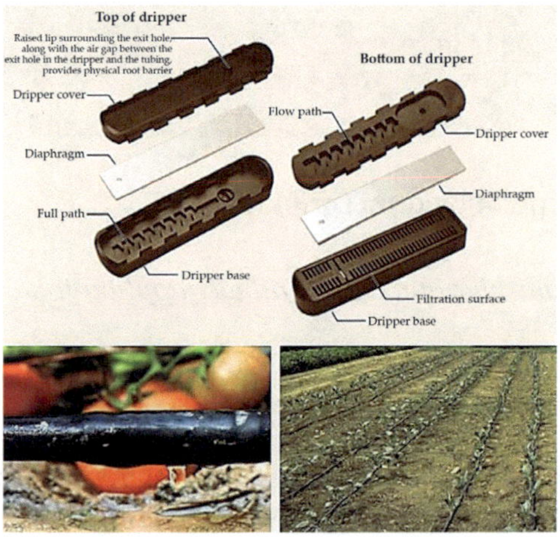

Figure 2.9: Dripper and drip line irrigation structure.

Source: https://www.semanticscholar.org/paper/Automation-i-63.-Automation-in-Agriculture-Edan-Han/774ae073f1ba426b147b4c0228e40b31b7cf08f0.

In sprinkler irrigation, water is supplied to numerous locations in the field and dispersed by high-pressure guns or sprinklers (Figure 2.7). Spatially variable irrigation structures had typically utilized self-propelled irrigation systems-sprinklers attached to center pivots or moving platforms (Klocke et al., 2003; Kim et al., 2006). Center-pivot irrigation is a sprinkler irrigation system that is made of several pipe sections merged that are attached on wheeled towers with sprinklers located along its length (Figure 2.8). The system travels in a circular array (Herwitz et al., 2004).

Drip irrigation systems (Figure 2.9) were created in Israel in 1965. Water is supplied slowly and properly to the soil, and only where required. A drip irrigation system contains of pressure regulators, backflow preventers, valves, emitters, filters, and of course the pipes (the mainline that takes water from the source to the valve, and the sub pipe that goes from the valves to the linking point of the drip tubes and the drip tubing). Low-head bubbler irrigation structures are micro-irrigation systems grounded on gravity flow that work at low pressure and needs no pumping or filtration. Their key advantages are lower energy requirements, rare mechanical breakdowns, and simplicity. But, their application is limited because of the installation problems and complex design (Yitayew et al., 1999; King et al., 2005).

2.3.2. Automation in Irrigation Systems

Automation systems comprise irrigation time clocks—electromechanical and mechanical timers to permit precise control of water replying to plant demands and environmental alterations, with recent progress in utilizing sensors to measure soil properties like salinity and moisture utilizing capacitance and resistance-based sensors, and time-domain reflectometry (Zazueta and Smajstrla, 1992; Cardenas-Lailhacar et al., 2005; Haley and Dukes, 2007). Sensors used for measuring plant stress through spotted and scanned canopy temperature measurements had been utilized in scheduling decisions for subsurface drip irrigation and center-pivot systems. Sensors comprise multispectral imaging, infrared thermometers, and thermal scanner sensors. (Wanjura et al., 2004; Evett et al., 2006) (Figure 2.10).

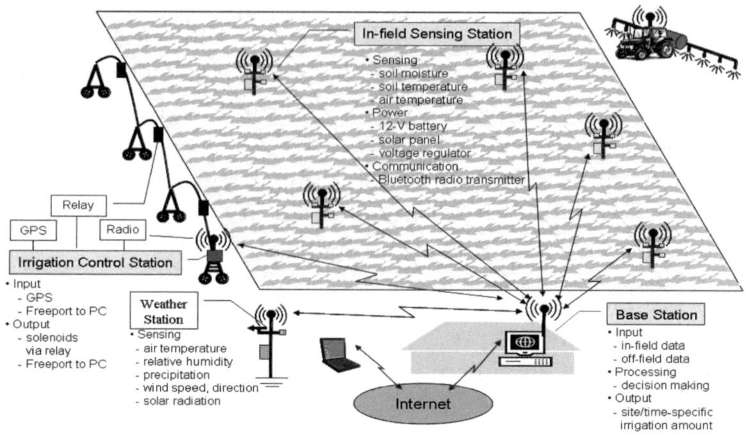

Figure 2.10: Wireless irrigation structure conceptual layout.

Source: https://www.researchgate.net/publication/226815941_Automation_in_Agriculture.

High-resolution information of water and soil dynamics joined with measurement of crop reply to water and salinity stress are significant for irrigation managing optimization. These data are usually delivered through weight-based soil lysimeters, with the latest development of a volumetric lysimeter structure (Poss et al., 2004; Raut and Shere, 2014).

Progress in automated irrigation systems comprise scheduling programs that utilize weather data to endorse and control time and crop growth stage, amount of irrigation, and nutrients /water requirements detected in real-time, and commercial produce monitors and remote sensors to chart crop production exactly. An instance comprises a real-time irrigation development program for additional irrigation that comprises an actual evapotranspiration model, a soil water balance model, irrigation forecast model, and a reference crop evapotranspiration model, all joined utilizing a mixed linear program (Upcraft et al., 1989; Hess, 1996).

Infrared sensor systems and a low-cost microprocessor for automating water penetration measurements are significant in controlling crop production and delivering agricultural chemicals and water to the soil profile. Control of nutrients with sensors allows fertilization management systems and optimization of irrigation, useful for decreasing environmental effect caused by the overflow of nutrients within surfaces and groundwater by utilizing ion-sensitive field-effect sensors (Artigas et al., 2001; Milla and Kish, 2006) (Figure 2.11).

Figure 2.11: Five in-field weather stations and sensing stations attached on the linear irrigation cart.

Source: https://link.springer.com/chapter/10.1007/978-3-319-01017-5_29.

A wireless in-field sensor-based irrigation management structure was developed to give variable-rate irrigation. Variable-rate irrigation was regulated by a computer that directs control signals to irrigation controllers through real-time wireless communications grounded on GPS positions of sprinklers and field information (Peters and Evett, 2005). A self-propelled linear sprinkler structure fitted with a program logic controller and a DGPS was distantly controlled through a base computer, utilizing a closed-loop irrigation system to describe the amount of irrigation created on dispersed soil water measurements. The system was functioned through a program logic controller that controlled solenoids to turn sprinkler nozzles off and on. The variable-rate application was applied by regulating pressure within a group of nozzles (King et al., 2000; de Miranda, 2003) (Figure 2.11).

To regulate small zones in field irrigation, micro-irrigation, and solid-set sprinkler could be controlled utilizing distributed or centralized irrigation controls (Miranda et al., 2005). Architectures of distributed sensor systems for site precise irrigation automation merging sprinkler valve controllers and smart soil moisture sensors had been developed and are commercially accessible (for example., Netafim, Irriwise). This could be extended to closed-loop control for automatic irrigation grounded on soil conditions and in-field sensing response of the plant. Additional developments comprise spot spraying of herbicides grounded on actual time weed detection utilizing optical sensors. Growers utilizing recirculating systems frequently choose to sterilize the drain water earlier supplying it back to the plants. One of the following two techniques is often utilized: ozone sterilization or ultraviolet (UV) sterilization. Automated real-time polymerase chain reaction (PCR) system for identifying pathogens in irrigation water had been created (Csordas et al., 2004; King et al., 2005).

2.4. GREENHOUSE AUTOMATION

The greenhouse environment is a comparatively relaxed environment for the introduction of automated machinery because of its structured nature. Therefore, the automated structure must deal merely with the variability of the agricultural produce. Hence, the development of a structure is simpler and easier. Automation structure for greenhouses deal with climate control, spraying, seedling, production, and harvesting as comprehensive are given in the following sections (Simonton, 1990; Pedersen et al., 2006).

2.4.1. Climate Control

Greenhouses had been developed in the twentieth century to retain solar radiation energy, to safeguard products from numerous harmful natural climates and insects, and to develop appropriate environments for plants by utilization of 2–3 mm glass plates or 100 μm plastic films. Developments in microcomputers and sensors had led to the latest greenhouse operations that comprise irrigation, nutrient supply to plants, and control of climate to produce the finest conditions for the growth of the crop in a reasonable way. Environmental control allows shorter cultivation periods and year-round culture. This segment outlines greenhouse environment automation and control (Kondo and Ting, 1998; Grift et al., 2008).

2.4.1.1. Sensors and Parameters for Environmental Control

Following are the constraints for environmental control (Monta, 1997a, b):

- CO_2 concentration;
- Airflow;
- Light; and

- Temperature and humidity.

2.4.1.2. Control Methods

Greenhouse climate control needs consideration of numerous nonlinear interconnected variables. Control models should be considered weather prediction models, the greenhouse model, and crop growth models. The following techniques had been utilized for control: artificial intelligence soft computing methods (neural networks, genetic algorithms, fuzzy control), classical methods (cascade control, proportional integral derivative control), and advanced control (predictive, adaptive, nonlinear) (Albright, 2001; Bailey, 2004). Control is applied with microcomputers or programmable logic controllers (PLCs). Climate controllers that utilize online measurements of the plant temperature, fruit growth, and quality, to approximate photosynthesis and actual transpiration would be the future progress. This would allow the progress of closed-loop systems that utilize the *speaking plant* as the response for the control system and thus result in ineffective regulation of the greenhouse climate. Effective regulation of the greenhouse climate must also include long-term management strategies to enhance quality and profitability (Serodio et al., 2001).

2.4.2. Seedling Production

Seedling production is one of the core technologies to produce high-quality products in vegetable and fruit production. Seedling operations like selection of seed, seeding, irrigating, transplanting, grafting, cutting, and sticking had been automated or mechanized (Kittas et al., 2001; Edan et al., 2009a, b). An entirely automatic seedling production factory had been reported as a portion of a plant factory; however, a specific seeding machine that can seed in a similar alignment had also been developed. Numerous robots for transplantation and grafting robots from cell tray to cell tray or pot had been commercialized (Van Henten, 1994; Weiss and Biber, 2011).

2.4.3. Automatic Sprayers

Chemical control is needed for crop production in precise environments, and automation of the chemical spray is required to reduce exposure of chemicals. Spraying robots had been commercialized so far (Gavric et al., 2011; Bac et al., 2014). A vital technology of the robots is independent control of the vehicle. The front axle of self-heading-correction machinery could be turned easily around an axis A-B which is static to the body diagonally. Supposing that a front-wheel on the single side runs on a ridge (it is of course), the midpoint of this front-wheel moves from O1 to O2. In the same period, the other front wheel changes down and back. Therefore, the subsequent steering angle β causes the vehicle to incline from the ridge, modifying its moving direction through itself. In the situation where both a rear wheel and a front-wheel run on a ridge at the same interval, the impact of the heading correction will be decreased due to the steering angle β might be smaller. To obtain a suitable steering angle, the rear tread is 35 millimeters smaller than the front tread (Caponetto et al., 2000; Morimoto and Hashimoto, 2000).

Another technique is called *electromagnetic induction type*. Induction wires are put down under headland and/or ridge aisles and a vehicle with an induction sensor recognizing the magnetic field produced by the wires can automatically move along the wires. When the vehicles travel to the next ridge aisle in thin headlands in greenhouses, numerous methods had been reported: four-wheel steering, a pivot shaft that comes out to take a turn, an extra rail system to carry the vehicle to the next aisle, and a manual technique. In orchards, automatic speed sprayers utilizing induction pipes and induction wires were developed in 1993 and 1994, respectively. A method that utilizes a remote-controlled helicopter had been very common in the fields (Arima et al., 1994a, b).

2.4.4. Fruit Harvesting Robots

It could be said that the history of agricultural robots initiated with a tomato harvesting robot (Kawamura et al., 1984). There had been considerable research on fruit harvesting robots for tomato, cucumber eggplant, strawberry, and cherry tomato (Kondo et al., 1993, 1996; Subrata et al., 1997). Vegetable harvesting robots had also been examined; however, there is no commercial robot yet. The key reasons restraining commercialization of harvesting robots are fewer success rates because of slow operational speeds, diversity of plant properties, and high costs related to the seasonal effect. Though, practical utilization of harvesting robots is probable in the future (Kondo and Ting, 1998).

2.4.4.1. Tomato Harvesting Robot

Investigation on the first tomato harvesting robot initiated at Kyoto University in 1982 and numerous different kinds of tomato harvesting robots and their constituents had been developed. A cluster harvesting robot is currently under development. The key components of numerous tomato harvesting robots are an end-effector, machine vision, manipulator, and traveling device, as shown in Figure 2.12. The robot automatically travels amongst ridges and breaks in front of a plant utilizing reflection plates and photosensors on the ridges, which could give the position of the robot in the greenhouse. When the traveling device rests, a machine-vision structure measures fruit color and location, the manipulator moves to the cluster, and an end-effector picks a fruit. After concluding the process at the location, the robot travels to the next location of the reflection plate.

Figure 2.12: A tomato harvesting robot.

Source: https://www.researchgate.net/publication/226815941_Automation_in_Agriculture.

2.4.4.2. Phytological Characteristics of Tomato Plant

Mostly tomato plants for the fresh marketplace are generally grown on a vertical plane with hanging equipment or with supports until numerous fruit clusters are harvested. Though, high-density single-truss tomato production systems (STTPS) had been reported (Ting et al., 1991; Kondo et al., 1993c). Also, an effort was done to grow the tomato plant upside down on the tomato production structure due to the smaller labor need for the comfort of mechanical operation and plant training. Some variations are for individual harvesting however, others are for cluster harvesting. Some variability produces longer fruits and round-shaped fruits, relying on the season. There are also numerous fruit sizes. Fruit clusters are assumed to grow outwards because of growth rule, however, the main stem occasionally twists, which caused random cluster direction hence tomato fruits might sometimes be secreted by stems and leaves. When a robot is presented to the production structure, it should be adjustable to plant diversity.

2.4.4.3. Manipulator

The basic process of a manipulator relies on the configuration of the plant, the three-dimensional (3-D) positions of its work items, and the method paths to the objects. In the first effort to robotize the tomato harvesting operation, a 5-DOF (degree-of-freedom) articulated manipulator was utilized, and a 7 DOF manipulator was examined for harvesting six clusters. Though the Dutch-style growing structure had been widely introduced to large-scale greenhouses all over the world, and target fruit is constantly located at the same height. Thus, a selective compliant robot arm (SCARA)-type manipulator could be used. When the fruit cluster is moved to a container rapidly, cluster swing damping is needed.

2.4.4.4. End-Effector

The fruit cluster had various fruits and their peduncles had joints in numerous varieties of tomato plants. When a person harvests ripe fruit one by one in the cluster, she/he could pick them off simply through bending them at the joints rather than cutting. To harvest the fruit, numerous end-effectors had been developed; Figure 2.13 shows one of them. A 10 millimeter-thick rubber pad is joined to every finger plate to shield the fruit from damage and slipping. The width, thickness, and length of a finger plate are 45, 10, and 155 mm, respectively. The gripping force applied by the finger plates could be adjusted from 0–33.3 N, however, these finger plates grip fruits extending from 50–90 mm in diameter. The suction pad was merged to the end of a rack, which is moved back and forth by a pinion and a DC motor amongst the finger plates. The stroke and speed of the suction pad motion are 80 mm and 38 mm, respectively. The suction pad could be moved forward up to 43 millimeters from the tips of the finger plates. The stopping position and moving distance of the pad could be detected through a rotary-type potentiometer. Two limit switches are joint to both ends of the pad stroke to avoid the pad from overrunning.

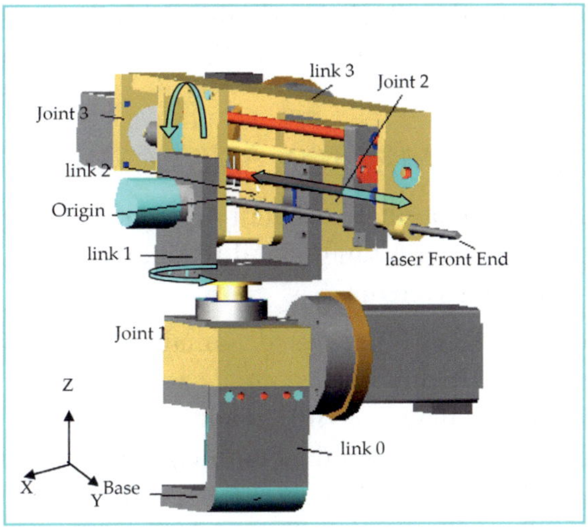

Figure 2.13: Diagram of end effector.

Source: https://www.researchgate.net/publication/299134929_Robotic_Delivery_of_Laser_for_Image_Guided_Orthopaedic_Surgery/figures?lo=1&utm_source=google&utm_medium=organic.

2.4.4.5. Machine Vision

A traditional technique of identifying 3-D locations of mark fruits is feature-based stereo vision. A pair of matching color cameras attain images and differentiate red-colored fruits. Based on the difference of fruits on both images, the deepness of the target fruit could be calculated. Though a little error in the 3-D location happens due to the hidden parts of the fruits, the suction pad could abide by these errors. It is not informal for stereo vision to identify all fruits locations when a corresponding difficulty happens because of hidden fruits and numerous fruits in the images. In this situation, an area-based stereo vision or 3-D laser sensor might assist to detect the fruit depths.

2.4.4.6. Traveling Device

Figure 2.12 displays a 4-wheel-kind battery car on which the tomato harvesting robot is attached. The traveling device travels and stops among the turns and ridges at the headlands to go to some other ridge. In the Dutch type of large-scale greenhouses, two heating pipes are normally used. It is very easy to familiarize a rail-kind traveling device as the pipes can be utilized as rails. Rail-kind traveling devices are already utilized for leaf picking, spraying, manual harvesting, and several other operations in greenhouses.

2.5. ANIMAL AUTOMATION SYSTEMS

Automation of systems of animal husbandry comprises the establishment of environmental control systems, automated feeding systems, and automated weighing and monitoring systems.

The controlling climate of the housed animals has a significant impact on the production and well-being of the animals and thus controlling it is very important. However, this may be a difficult and problematic task because of the nonlinear effects of animals on the humidity and temperature conditions inside the animal housing buildings. In conditions where the animals are kept outside, climate control is further intricate because of changing environmental conditions. Environmental and air-quality monitoring is significant for environmental protection features and therefore is gaining increasing importance and attention (Eradus and Jansen, 1999).

The device for electronic identification and monitoring of animals became accessible in the mid-1970s and has empowered execution of advanced management schemes, explicitly for livestock and the swine management. The ISO regulation of injectable electronic transponder in the late 1990s extended applications to all of the animal species. Various sensors have been made to deliver individual animal parameters like size, fat, and weight. These parameters are utilized for management decisions. The fresh new generation of the sensors assists production and health status monitoring, both refining animal welfare and confirming increased food safety and quality. Recent developments comprise acoustic passive incorporated transponder tags using micro-electromechanical systems (MEMS) technology. Tags may be utilized for locating animals from growth to the final processing for food security purposes and quality control (Daskalov et al., 2006; Holm et al., 2007).

The main expenditure in the animal production systems is the intake of food. Automated feeding systems reduce production costs whereas confirming that animals receive essential nutrient ingredients. Individual and group feeding systems have been made to measure and manage food intake.

Production, welfare, and health controls are being familiarized into modern farms utilizing advanced information systems. Data from several sensors at the group and individual levels are taken daily for advanced control and monitoring. Several systems will be defined in the subsequent sections.

2.5.1. Dairy

The dairy industry is possibly the utmost automated agricultural production system, with nearly all procedures, from feeding to milking, being fully automated. In the dairy industry, numerous maintenance routines like milking, weighing, feeding, and online record of performance are completely automated on a specific animal basis. Optimal management is described as generating maximum milk yield while reducing costs. The computation and storage of the data capacity of computers ideally empower sophisticated decision-making to support the automated procedures to obtain optimum individual and herd performance. These comprise automated feeders, sensors that compute daily activities of cows and online automated shop systems for recording production and quality of milk. Reproduction monitoring comprises systems for the timing of insemination centered on oestrus detection. Health care systems comprise the detection of mastitis. The aim is to completely automate every procedure from feeding to milking to decrease production expenses and maximize the yield of milk.

The physical procedure of feeding and recording real feed ingestion is centered on feed management of roughage and concentrates, ration composition, and feed computation for a group of cows or individual cows. Analysis of the performance data specifies that the performance of cow under the uniform rationing rule is steady in trend but changes in magnitude, and thus an optimum feed policy, in terms of *effective* rationing of concentrates, must be on the individual basis (Maltz et al., 1992). An alternative method, the sweeping technique, is centered on the average values for the herd. This can trigger cows not to obtain maximum milk yield due to inadequate concentrate ration or suggest that surplus feed be consumed as there exist cows that might have reached the maximum yield of milk with the smaller concentrate ration. Both outcomes in redundant financial expenses. Because of the arrival of technology, the farmer is capable to allot a different quantity to each cow utilizing individual computer-managed calf feeders and incorporated real-time control systems for controlling, measuring, and monitoring individual intake of food of the free-housed dairy cows (Halachmi et al., 1998; Seipelt et al., 2003).

Individual distribution decisions are made according to the performance of each cow, Performance parameters comprise the output of individual cow and measurements of the physiological variables comprising body composition, size, and shape. An instance of a system comprising of forty feeding cells is exhibited in Figure. 2.14. Each cell consists of an identification system, automatic closing, and opening and yoke gate and a fodder weight system (Fisher, 1997). Each feeding stall comprises a feeding trough, CPU (central processing unit) and an electronic weight scale, identification system, cylinder with valve, and a presence sensor. All components are attached to a PLC which processes the data and triggers the electro-pneumatic actuators. The data is generally backed up to the management computer. The management computer is utilized as the monitoring station and the basic man-machine interface for describing basic operations and initial data analysis. The precise yoke design permits the head of the cow to enter the gate of yoke without allowing access to the fodder. This places the RF (radio frequency) identification tag on the ear of a cow close to the antenna and instantaneously activates the proximity sensor. If the cow is permitted to eat according to the programmed conditions, the PLC records the existing scale's weight and the gate of the yoke bar is lowered by the linked electro-pneumatic cylinder. The cow might then drive its head into the fodder and feed. The scale computes and records the weight of fodder at programmed intervals. Each scale of CPU is joined to the PLC directly through binary code to decimal (BCD) so no delay of time is caused by the weight transmission. The restriction bar on fodder trough averts the cow from pushing the head up, thus averting spillage of fodder. The utilization of a presence sensor in the yoke seemed to be very significant to conclude if the cow is not in the yoke station. The feeding troughs were prepared in a row to facilitate suitable dispersal of fodder by the path of the semi-automated fodder dispersal wagon (Filby et al., 1979).

Figure 2.14: Controlled automatic fodder ingestion and feeding system.

Source: https://www.semanticscholar.org/paper/Automation-i-63.-Automation-in-Agriculture-Edan-Han/774ae073f1ba426b147b4c0228e40b31b7cf08f0.

Various methods have been made for the automatic weighing of the cows. Cows are normally weighted as they leave the milking parlor to not disturb their daily regime. The motion of cows causes measurement problems, comprising variations along the scale because of applied forces, crowding of the cows on scales, and vital changes between cows and amongst the same cow at several times of a day or on different days. Vibrant weighing of cows is generally a common practice in several commercial farms, accomplished by cleaning the measured signal and then averaging it or recording its peak value as the cows transfer their weight utilizing physical-mathematical simulations that simulate cow walking (Ren et al., 1992; Peiper et al., 1993; Cveticanin and Wendl, 2004).

Milking cows is an intricate task because of the physics combined and variable biological components comprising the hazard of contaminating the udder with pathogen microbes (Ordolff, 2001). Even though the first suggestions for mechanical milking were given over a hundred years ago, the machinery of milking became common in the early 1950s, with fully automatic milking systems being presented in the 1990s.

First steps in computerizing the milking process comprised recognition of the end of milking and computerized teat cup detaching. Several optical, inductive, and capacitive sensors were made to identify low milk flow, which specified the end of milking (Ordolff, 2001; Juszka et al., 2011).

Mechanized stimulation of the udder was accomplished by utilizing pneumatic and electronic pulsators. Continuous individual changes in pulse rate, vacuum level, and pulse rate for every milking unit were made. Milk produce recording is executed using tipping trays and the volumetric measuring systems, with numerous sophisticated measuring systems to isolate air from milk to enhance accuracy. Automatic milking needs an automatic application of the teacups. Ultrasonic sensors, charge-coupled device (CCD) camera, and a laser are utilized to track the teats to manage in real-time the arm to familiarize with the changes in spacing's, teat positions, and shape and to motions of the cows during teat attachment. In most of the systems, a 2-stage teat location procedure has been applied.

First, the estimated teat locations are determined through dead reckoning utilizing vision systems or ultrasonic proximity sensors, body position sensors. The last attachment is attained by fine-position sensors utilizing arrays of light rays fixed on the robot arm. Automatic checks of milk quality and udder condition comprise online milk analysis. Milk quality is a crucial parameter both from health perspectives and from an economic point of view. Measures contain the color of milk, temperature, conductivity, incorporated with yield information. Biosensors had been utilized to measure metabolites, mammary infection components, and antibiotic residues, comprising the development of electronic analysts that assist real-time measurements. Online aligned milk composition sensors calculate in real-time in milking the concentrations of protein, lactose, and fat and indicate the existence of somatic cell count (SCC) and blood grounded on near-infrared (NIR) analysis (Jenkins et al., 2002; Katz et al., 2007).

Several teat cleaning systems, comprising rollers and brushes or isolated teat-cup-like cleaning equipment, had been developed. Furthermore, systems for cleaning the whole system (cleaning with boiling water, cluster flushing, and circulation cleaning) are applied.

Robot milking initiated in the early 1990s by numerous commercial companies (for example, GM Zenith, Fullwood Merlin, Lely), provides enhance yield by growing the frequency of milking and enhanced milk quality (Figure 2.15).

Figure 2.15: Lely milking robot in an open shed with fans controlled when cow gathering is detected.

Source: https://www.semanticscholar.org/paper/Automation-i-63.-Automation-in-Agriculture-Edan-Han/774ae073f1ba426b147b4c0228e40b31b7cf08f0.

Automatic health measurements throughout automatic milking comprise respiration rate measurement and leg health measurement (Rajkondawar et al., 2002; Pastell et al., 2006). Lameness detection is significant due to the key health, economic, and welfare problems it causes. Leg health is similarly measured through measuring the load of each leg or dynamic weight however the cows are weighed on weighing machine at the withdrawal of the milking parlor. Several methods had been developed comprising force plates that calculate reaction forces on a walk through weighing structures, activity meters worn around the neck, and enhanced respiration rate measured utilizing laser distance sensor. Ultrasonic back-fat sensors could provide information about the growth or health status of the livestock (Tasch and Rajkondawar, 2004; Pastell et al., 2006).

Another significant health measure is mastitis, a key reason for early losses in cows, and decreased milk yield, caused through the biological activity of microbes. It could be detected by calculating the quantity of somatic cells in milk. Several methods had been developed to measure it precisely utilizing body temperature, milk temperature, and electrical conductivity measurements. Lately, inline NIR sensors had been developed to measure milk temperature and milk conductivity of every separate quarter (a sensor is linked to every udder cup) (Frost et al., 1997).

The main direct parameter to identify oestrus is the concentration of milk hormones (progesterone), signifying the fertility status of the cow. Though, it is usually measured merely in laboratories based on samples, though biosensors had been developed for its measurement (Delwiche et al., 2001; Gygax et al., 2007). Numerous indirect parameters had been developed into automated structures, comprising milk temperature, the electrical conductivity of vaginal secretion, and cow behavior comprising cow activity measurement utilizing pedometers, heart rate, etc. Improved measurement was attained through combining information from numerous parameters, for example, feed intake, and milk temperature combining cow activity with milk yield).

Behavior measurement had been attained by utilizing different structures: a radar-based automatic native position measurement structure for tracking dairy cows in free-stall sheds, video measurements, global positioning structures for measuring grazing attitude (Turner et al. (2001) and automatic tracking structures based on the magnetic induction (DeVries et al., 2005).

Environmental regulatory systems in the dairy industry are fewer common as cows are positioned in barns that are open, partially shaded or shaded systems developed comprise automatic cooling utilizing fans based on online imaging structures that detect microclimate, gas emissions and crowding in cold uninsulated cattle farms (Teye et al., 2006; Huhtala et al., 2007).

Management information systems that associate individual health and herd and production parameters are significant to ensure effective automation (Baars et al., 1996; Van Asseldonk et al., 1999). Further developments in the design and supervision of livestock environments would need the development of maintainable livestock production structures accounting methodically for the environmental welfares and burdens of the procedures utilizing a lifecycle assessment procedure. Strategies will be required to be developed to control and decrease harmful gas emissions from land application of manure and livestock farms (Tomaszewski et al., 2000; Powell et al., 2007).

2.5.2. Aquaculture

Physiological amounts of cultured species could be regulated through controlling the system inputs and environmental conditions. This produces enhance process effectiveness, decrease energy and water losses, decrease stress, decrease labor costs, and disease. Automation applications comprise algae and feed management, feed production, environmental controls like automated air-pressure control and filtration systems. A severe water quality checking program comprises routine sampling (twice a week) and 24-hour sampling (every three months) of flow rate is regulated through regulating air blower frequency utilizing measuring of oxygen concentration in the fish container nitrogen (NO_2, NO_3), Ph, temperature, and phosphate. Shellfish and fish biomass should be tested and seaweed should be harvested (Mozes et al., 2003; Martins et al., 2010).

Automation generally exists in closed structures like recirculated aquaculture structures, but it could also be applied to offshore and pond aquaculture systems. Intensive recirculating aquaculture systems (RAS) decrease water and land utilization at the expense of enhanced energy need for operating treatment procedures to assist high culture densities, frequently with the addition of clean oxygen. The utilization of pure oxygen is normally expensive and needs substantial energy for dissolving in the water also for stripping off the carbon dioxide produced by respiration. In conventional RAS design, gas interchange and dissolved leftover treatments (for example, ammonia removal and CO_2 stripping by nitrification) are connected into one water treatment loop. Though, because excretion amount of CO_2 are an order of magnitude more than ammonia excretion amount this design might consequence in poisonous CO_2 concentrations. In addition, pure oxygen addition and pressurized pumping might enhance the danger of gas bubble disease. Therefore, low-head recirculating structure that detached the gas treatment loop (CO_2 and oxygen) from the solid filtration and nitrification treatment loop by utilizing a high efficient airlift generating a bubbly flow are utilized (Blancheton et al., 2006). The integrated pond system (IPS) concept recommends a novel solution for environmentally welcoming land-based mariculture. The IPS recycles expelled nutrients (valuable nitrogen) by algal biofilters using solar radiation for their photosynthetic procedures (Shpigel et al., 1993).

Accurate shape and size information of cultured and wild fish population is significant for organizing the growth and harvesting procedure comprising grading times, feeding regimes, and optimal harvest time (Lines et al., 2001; Costa et al., 2006). Information on both distribution and average weight is essential for feeding, grading, and harvesting results. Machine vision had been utilized to determine the fish size, weight, color, and activity outlines. The difficulties with images taken in ponds are the low contrast amongst fish, changing lighting conditions, and the dynamic movement of fish. Actual time in situ fish biomass estimation and behavior quantification had also been utilized for management decisions (Lee, 1995; Zion et al., 2007).

The cost of feed is generally the main operating cost in aquaculture. Overfeeding consequences in leftovers, which leads not only to poor water quality and also to extra costs, causing extra loads and additional stress on bio-filters, oxygenation, and mechanical devices (Chang et al., 2005).

Furthermore, feeding rhythms upset the proximal structure of fish flesh and feed conversion rates. Automated feeding structures comprise demand feeders, timer-controlled feeders, and automated data-acquisition structures to evaluate acoustic and fish feeding rhythm, photoelectric sensors to identify the turbidity of the discharge. Machine-vision structures and hydroacoustic sensors had been utilized to identify left-over pellets (Papandroulakis et al., 2002) (Figure 2.16).

Figure 2.16: Aquaculture closed structure with a feeding station. Feed quantities are measured on a daily basis to every fish tank according to water temperature, fish weight, and growth rate.

Source: https://www.semanticscholar.org/paper/Automation-i-63.-Automation-in-Agriculture-Edan-Han/774ae073f1ba426b147b4c0228e40b31b7cf08f0.

Future research should be made towards controlling recirculated systems, the development of sustainable automated structures, and engineering environmental monitoring. Considering that maintainable development is perhaps the main challenge met by aquaculture (Muir et al., 1999), one must consider sustainability, which could be measured in three major classes: economical, sociological, and environmental (Caffey, 1998; Wurts, 2000). Another view of sustainable development associates to external effects and resource utilization that are described through various indicators (generally in physical terms). This should comprise automation of the final harvesting, online reporting of structure failures, and grading procedure, thus maintaining product quality and improving food safety (Wheaton and Hall, 2007).

2.5.3. Poultry

Poultry house controllers comprise sensors for external and internal temperature measurement, static pressure, water consumption, moisture, gas, and vent box status, static pressure, and feed lines. Additional automation machines comprise feed consumption checking equipment, feed bin load sensors, bird weight scales, gas meters, and water meters (Figueiredo et al., 2003; 2004).

Physiological signals are significant for behavior analysis and health monitoring. Numerous structures had been developed, comprising an implanted radio telemetry structure for remote monitoring of deep body temperature and heart rate and multispectral image examination for real-time disease finding. An automated nutrition and growth control structure had been developed for broiler creation utilizing an online parameter estimation process to model the vigorous growth of broiler chickens as a reaction to feed supply (Sergeant et al., 1998; Chao et al., 2001). Image-based bird behavior examination could be utilized to develop time outlines of bird activity (response to ventilation, movement, huddling, etc.) also to compare activity levels in diverse portions of the house. Time outlines of bird activity could contribute to improved water design and feeder, and improved distribution of ventilation air to deliver more uniform bird comfort (Rabatel, 1988; Jaiswal et al., 2005).

Numerous mechanical poultry catching structures had led to increases in bird welfare in adding to a decrease in manual labor. Systems comprise: rubber paddles that rotate on the birds from above and then move the birds onto a conveyor belt which transfers them back to a loading platform where they are placed into crates, a hydraulic drive structure that spreads along the poultry house and lift up the birds with soft rubber fingered cylinders that softly put them onto a conveyor that relocate the birds to an Anglia Auto flow and a caging system (Norfolk, UK) batch mode catcher that transfer birds from collection to a separate packing unit (Trevelyan, 1989; Yamashita and Kondo, 1992).

2.5.4. Sheep and Swine

Robot shearing operations had been developed and commercially functional in Australia. The sheep is controlled with straps on a portable platform. Hydraulically position clippers utilizing force feedback regulate the actual clipping of the wool. Path computations are constantly updated throughout the shearing process (Perez-Munoz et al., 1998).

Numerous feeding systems occur in sow farming: commercial electronic feeding structures that feed one at a time through enclosing every sow as it eats, electronic sow feeding structures in loose housing environments that restrict the feed ration, and a computer-controlled structure that permits sows to feed from 1/2 feed formulations to fulfill their nutritional need while satisfying their requirement for satiety by utilizing bulk ingredients giving automatic average daily weight gain and body weight (Lopes et al., 2002; Wang et al., 2006). A significant indicator of animal health and growth is the animals' weight, in adding to its significance in determining eagerness for market. Weighing had been accomplished utilizing walk-through weighing grounded on imaging systems and mechanical scales (Schofield et al., 2002). Physiological variables measurements comprise body size and shape utilizing image analysis. Ultrasonic probes had been applied to determine backfat for checking feeding regimes and animal growth. A robotic structure capable of carrying a sensor and placing it on the pig, however, it is positioned in the feeding stall had been developed (Tillett et al., 2002; Xin and Shao, 2005).

Regulating of swine thermal ease and actual time behavior had been attained utilizing imaging systems (Barbari, 2005). Planning individual showering structures for pregnant sows to avoid heat stress has been utilized in automatic shower cages to avoid wastewater and increase efficiency. Automatic cleaning structures to decrease infections risks amongst batches of pigs had been utilized based on an intelligent sensor for robotic vacuuming (Saeys et al., 2005; Zhang et al., 2006).

Latest environmental policies restraining the amount of phosphorus and nitrogen that could be applied in the field had led to the development of online examination of pig manure structures, comprising mobile spectroscopy devices in the visible and near-infrared wavebands (Harrell et al., 1990; Hoyt Jr et al., 2007; De Villiers, 2008).

2.6. FRUIT PRODUCTION OPERATIONS

Fruit production automated structures deal with all phases of production: growing (weeders, automated sprayers), harvesting, and post-harvest operation (sorting, grading).

2.6.1. Orchard Automation Systems

Fruit production procedures in orchards like pruning, thinning, spraying, harvesting, and weeding had been automated and mechanized. Even when automation structures had been developed for a similar variety of fruit tree, their constituents differ significantly due to plant training systems, climate conditions, cultivation methods, labor conditions, and other situations and conditions vary from country to country. This section describes mechanisms, functions, and significant observations of automation structures in orchards (Tanigaki et al., 2008; Sears, 2015).

2.6.1.1. Fruit Harvesting Robots in Orchards

Numerous kinds of shakers are operational in orange fruit orchards: canopy shake and catch, constant canopy shake, mono-boom trunk shake, trunk shake and catch, and others. These shakers are utilized because of labor shortage, however, the harvested fruits are merely for processing into juice; they can't be used in the fresh market due to unavoidable damage. Numerous kinds of orange harvesting robots that had manipulators with picking end-effectors and machine-vision structures had been reported in Japan, the USA, Japan, and EU countries (Juste and Fornes, 1990; Kondo, 1994; Rabatel et al., 1995).

It was made by Kubota Co., Ltd., Japan. The benefit of the articulated manipulator is its compressed size when folded up in a slim space amongst trees. Citrus trees have huge canopies and many branches, leaves, and twigs. Since these can be often difficulties for fruit harvesting, research on robots with more grades of freedom had also been reported (Moltó et al., 1992; Hannan and Burks, 2004). Color cameras were often utilized as sensing structures to identify fruit because citrus fruit have orange colors. Fruit places are calculated by utilization of stereo vision, vision servoing (vision-based robotic control), ultrasonic sensors, or a mixture of them and differential object size (Slaughter and Harrell, 1987; Kondo, 1995). Their end-effectors had the job of rotating semicircular cutters so that they could cut peduncles in numerous directions (Sittichareonchai et al., 1989; Edan and Miles, 1993).

2.6.1.2. Automation of Spraying and Weeding Operations

Control of disease, weeds, and insect pests is a vital operation to achieve a stable high produce of crops and high-quality products. This operation comprises chemical, physical, and biological methods. Chemical spraying is extensively utilized in agricultural production environments. Nowadays, technologies with high exactness to spray only the essential parts of the plant utilizing a minimum amount of chemicals are needed to protect the environment as well as workers (Grand D'Esnon, 1984; Kassay, 1992).

A nozzle-positioning structure for an exact sprayer was studied with a vigorous crop position revealing system. The information from a vision sensor was conveyed to a herbicide applicator that is produced up of a microcontroller, with slidable arms attached with spray nozzles. Nozzles were driven to the optimum positions. The system was verified to assess its performance. It had high sufficient accuracy for utilization in Japanese rice crop fields. A fluid handling structure to permit an on-demand chemical injection was made for a machine-vision-controlled sprayer. The structure was able to deliver an extensive range of flow rates of a chemical solution (Rabatel, 1988; P. Gillis and K. Giles, 2001).

Wiedemann et al. (2002) created a spray boom that could touch mesquite plants. Sprayers were fitted to all-terrain vehicles and tractors. Controllers were designed to direct stable duration voltage pulses to solenoid valves for release of spray by flat-fan nozzles when mesquite canopies obstruct the light. The stages of mesquite mortality attained were equivalent to those attained by hand-spraying through ground crews (Iida et al., 1996; Hayashi et al., 2005).

The speed-sprayer (SS) had been extensively utilized in orchards and its independent control is a key theme in the automation of the spraying process. The *pipe induction type and electromagnetic induction type* were commercialized in 1994 and 1993 (Cho, and Lee, 2000; Hadlow, 2016); both types need induction pipes or wires in the underground, on-ground, or above ground at 150 to 200 cm height and between tree rows. Safety sensors (touch sensors, ultrasonic sensors), induction sensors, and other internal sensors are fitted in the unmanned SS, and independent control is led with fuzzy theory. Another technique of SS control with genetic algorithm and fuzzy theory utilizing GPS had been reported (Dohi, 1996; Ahmad et al., 1998, 1999).

When an end-effector is joined to the manipulator, it could weed on the ridge amongst crops. Color images of the weed are input to a computer through a color camera and the 3-D location of the weed is measured utilizing a binocular stereo method. Weed detection is led utilizing texture or color difference amongst crops or soil and weed the end-effector was a weed knife with a curved shape (4 cm diameter). This robot could also be a transplanting robot or a leaf vegetable harvesting robot when the end-effectors were substituted (Kurokami, 1980; Nishiwaki et al., 2004).

2.6.2. Automation of Fruit Grading and Sorting

Due to the ever-growing requirement to supply high standard food products in a short time, automated categorizing of agricultural products is receiving special priority amongst numerous farmers associations. The motivation for these trends could be credited to increased knowledge of consumers regarding their health well-being and a reply from producers to deliver quality guaranteed products with constancy. It is in this perspective that the field of machine vision and automatic inspection comes to play the significant role of quality regulator for agricultural products. Unlike many industrial products, quality review of agricultural products presents precise challenges due to nonstandard products must be examined according to their internal quality and appearance, which are suitable to customers only for nondestructive techniques. Numerous sensors have been made and applied for internal quality checking, comprising acidity, sugar content, rotten core, rind puffing, and other internal defects (Steward et al., 2002) (Figure 2.17).

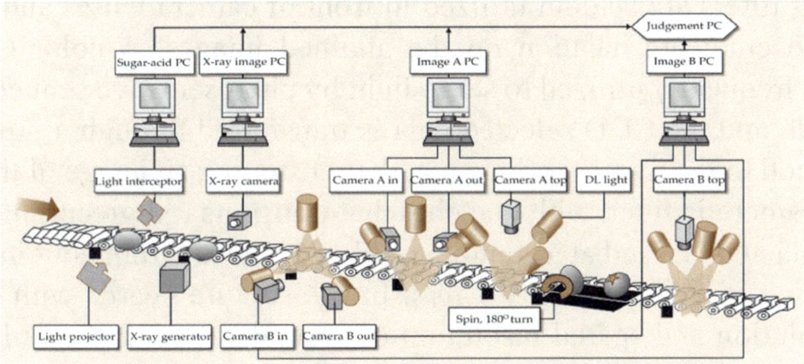

Figure 2.17: Schematic illustration of lighting and camera setup.

Source: https://www.semanticscholar.org/paper/Automation-i-63.-Automation-in-Agriculture-Edan-Han/774ae073f1ba426b147b4c0228e40b31b7cf08f0.

2.6.2.1. Fruit Grading System with Conveyors

As a demonstrative of other agricultural produces, discussion in this segment is focused primarily on the orange fruit, a main agricultural product examined by this system. The key components of the system for sorting and an automated inspection could be outlined as follows:

- Product reception from seller;
- Container emptying and discarding of products;
- Feeding of products to the conveyor line;
- Inspection for external and internal situations and defects followed through the assignment of quality rating;
- Adjustment of weight and release of the examined product into packing box;
- Labeling of size and grade utilizing an inkjet printer;
- Box closing and sealing; and
- Box transfer on palette and loading completed for marketing.

2.6.2.2. Illumination and Image-Capture Devices

Illumination is one of the most significant parts for the machine vision system to examine products, because it decides the quality of images attained, particularly for glossy produces whose cuticular coatings are thick. Polarizing filters are seldom utilized in front of camera lenses and lighting devices to eradicate halation on the attained images. A color CCD TV camera is frequently utilized to sense light by photosensitive semiconductor equipment, and the CCD selected data is transferred through a progressive scan method to the frame storing area, demonstrating an image of the scene. The TV camera is fitted with one chip that transmits red-green-blue (RGB) analog data and is fixed at a shutter speed of 1/1000s throughout inspection because line speed is typically 1 m/s. Image capture boards with eight-bit level resolution and spatial resolution of around 512×512 pixels had been utilized to store and digitize video signals and output the information to computer memory for examination or show on the monitor. Lately, an exceptional image-acquisition device, a universal serial bus (USB) or a local-area network (LAN) is often utilized amongst TV camera and personal computer (PC) rather of the image capture board, allowing image processing to be achieved in 10 to 30 ms (Figure 2.17).

2.6.2.3. Product Reception and Forwarding

The initial step in the inspection process starts at the delivery platform, located on the ground floor. Agricultural products are packed in the containers by the farmers and supplied to the examination factory in trucks. A folk lift is utilized to unload the containers on a single pallet and supply them to the depalletizer device that splits the containers automatically so that they are served one by one to the conveyor, which pushes them to the upper floor, where the key inspection line is situated.

2.6.2.4. Internal Quality Inspection

The first phase in the inspection series is to determine the acid and sugar contents utilizing a NIR inspection system. A distinctive sensor decides the acidity level and sugar content (brix equivalent) of the fruits from light wavelengths obtained by precise sensors after light is transferred through the fruit. The sensor photo-electrically alters the light into signals and directs them to the computer unit, where they are treated and categorized.

2.6.2.5. Image Analysis

At the next process, the fruit is carried to the third inspection phase where the key image processing and grading took place utilizing factory automation computers. Six CCD cameras fixed in random trigger way obtain images of the fruits as they are carried at a constant speed. Initially, two side cameras on the right side, positioned equidistance from the central point of the fruit, take images of the right side of the fruit. Next, two side cameras on the left side, again positioned equidistance from the central point of the fruit, take images of the left side of the fruit. Next, a top camera obtains the top surface image. Lastly, the fruit is then rotated through 180° round a horizontal axis through mechanically controlled roller pins, which enables the attainment of an additional image of the lower surface through a sixth camera. Images are captured through CCD cameras after a trigger signal is obtained by the DIO (digital input/output) board. Sensors wired to a sequencer box, are utilized to track the fruit position and the output is conveyed to the DIO board from where the program studies the 8-bit output data.

Image processing is then performed and the following features are examined:

- Shape (inflection point of the external surface, the center of gravity, L^2/area, perimeter L);

- Size (minimum and maximum diameter, extrapolated diameter and area);
- Bruise (based on the summation of color levels for blue and intensity of blue color level and R-G derived images); and
- Color (color space based on saturation, RGB ratios, hue, intensity (HSI) values).

2.6.2.6. Data Maintenance

A significant feature of the grading structure design is that it is adjustable to the inspection of diverse products like persimmon, sweet pepper, tomato, potato, kiwi fruits, and waxed apple with adjustments merely to the processing codes. Numerous lines for orange fruit examination combined with high-speed microprocessors and high conveyance permit the system to control large bunches of fruit products at high speeds. Apart from the quality examination, another objective of this structure is to gather data regarding product performance. Identification of certain counts and flaws could lead to the discovery of the source and its severity. All information from receipt of fruits at the collection place by the analytical procedures, packing, and shipping is kept in an office computer connected by a local-area network (LAN) to the following factory automation PCs. Experienced employees manage the data, grounded on which day-to-day actions could be monitored remotely. By creating product performance records accessible on the Internet, it will be probable to monitor performance online and provides a valuable service to the customers. All in all, the client of a product is the final evaluator of its quality. Thus, keeping specifications and internal standards in line with customer prospects is a precedence that is achieved by good relationships and frequent communications with the customer.

2.6.2.7. Fruit Grading Robot

Established on the technologies of the grading structure described in the previous segment and robotic technologies, a fruit grading robot structure has two three-DOF Cartesian manipulators, 16 machine-vision systems comprising of 16 color TV cameras, 16 suction cups as end-effectors, and 36 lighting devices. This could be applied to peach, apple, pear, tomato, and other fruits (Figure 2.18).

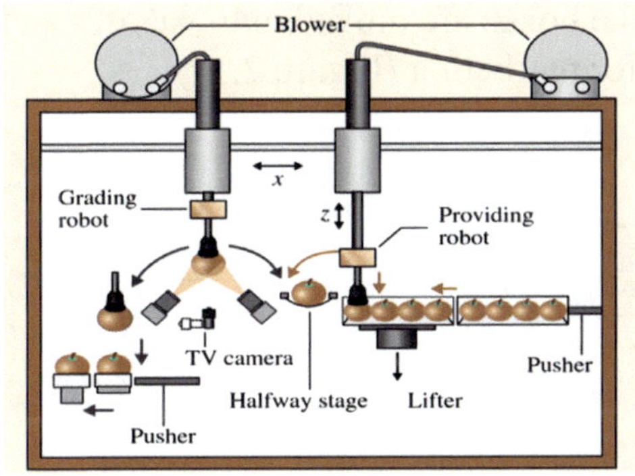

Figure 2.18: Robotic actions.

Source: https://www.semanticscholar.org/paper/Automation-i-63.-Automation-in-Agriculture-Edan-Han/774ae073f1ba426b147b4c0228e40b31b7cf08f0.

Operation flow with this grading structure is as follows:

1. When products are obtained, 4 blocks of containers (a block contains 10 containers) are burdened on a pallet;
2. A block of the containers is raised up to the second floor, where the key parts of the grading system function and a container divider forward containers one by one to a barcode reader;
3. After attaining information from every barcode attached to a container, the container is forwarded to a robot, the fruit providing robot;
4. The robot pulls fruits up by utilizing suction pads and transfers them to a halfway stage;
5. Another robot, the grading robot, lift fruits up once more from the halfway stage and side and bottom images of fruits are attained by TV cameras through transferring fruits;
6. The robot supplies them to trays on a conveyor line and an upper image of the fruit is attained in a camera box;
7. After appearance examination, sugar content and internal conditions are examined by an infrared analysis sensor;
8. Fruit that passes through the internal quality sensor box are packed into a corrugated cardboard box through a packing robot based on their grading outcomes. Name, size, and Grade of fruit variety are written on the box surface through an inkjet printer and the box is closed and sealed; and

9. Lastly, the boxes are moved onto palettes and are burdened into a truck for marketing (Figure 2.19).

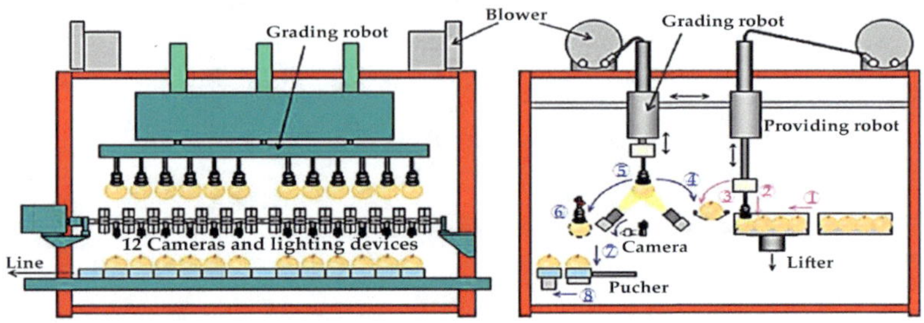

Figure 2.19: A tomato fruit grading robot.

Source: https://www.researchgate.net/figure/shows-a-fruit-grading-robot-system-installed-at-JA-Shimoina-Japan-The-robot-system_fig11_270956937.

Figure 2.19 shows the activities of the two manipulators: the grading robots and fruit providing robots. A container in which 6×5, 6×4, 5×3, and 8×6 fruits are filled is pressed in the working area of the providing robot through a pusher (1). The providing robot had a 3-DOF Cartesian coordinate manipulator and 8 suction pads as end-effectors. The robot pulls 8 (maximum) fruits up (2) and transmits them to a halfway stage, lay-outing fruit intervals in the *y*-direction (3). 2 providing robots autonomously work and set 16 (maximum) fruits on a halfway stage. A grading robot which comprises of another 3-DOF manipulator (a rotational joint and two prismatic joints) and 16 suction pads pulls them up again (4) and transfers them to trays on a conveyor line. Bottom images of fruits are attained as the grading robot moves over 16 color TV cameras. The lighting devices and cameras turn down 90◦ subsequent the grading robot's motion (5). Before discharging the fruits to trays on the line, the TV cameras obtain 4 side images of fruits as they rotate by 270◦ (6). After the image attainment, the robot discharges the fruits into trays (7) and a pusher pushes 16 trays to a conveyor line.

In developed countries, the production preference is towards huge scale farms and therefore automation will be commercialized and advanced to make this possible. Farmers must yield their food at economical prices to stay in business and computerization of farming technology is the only way for advancing. With the enhancement of computers and sensors, and reduction of automation machinery costs, this is becoming possible and more systems will be presented. The existing century will perhaps see substantial advances in robotization and automation of farm operations. The future farm will comprise the incorporation of advanced sensors, intelligent software, controls, to give viable solutions to the complicated agricultural environment data of every fruit is sent from a computer. This grading robot's extreme speed was 1 m/s and its stroke was around 1.2 m. It took 2.7 seconds for the robot to hand over the 16 fruit to trays, 0.4 seconds to move down from the primary position, 1 second to move back from releasing fruits, and 0.15 second for waiting. The total time was 4.25 seconds to move back and forth for the blow. This makes this robot performance nearly 10000 fruit/hour. In this structure, 4 blowers with the specification of 3400 rpm, 1.3 m^3/min displacement, 38 kPa, 1.4 kW were utilized for 2 providing robots and a grading robot. Around 30 kPa vacuum forces was appropriate for sucking peach fruit, however, 45 kPa was utilized for pear and apple fruits and no harm was detected however after sucking peach fruits twice. The tray had a data carrier (256-byte EE-PROM), and grading low-cost labor in third-world countries, considerable work on the farms is yet performed manually. Regardless of the large capital investment required to purchase the equipment, automation will possibly be introduced also in these countries to deliver the needs for land efficiency and enhanced production (Njoroge et al., 2002).

REFERENCES

1. Ahmad, U., Kondo, N., Arima, S., Monta, M., & Mohri, K., (1998). Weed detection in lawn field based on gray-scale uniformity. *Environment Control in Biology, 36*(4), 227–237.
2. Ahmad, U., Kondo, N., Arima, S., Monta, M., & Mohri, K., (1999). Weed detection in lawn field using machine vision. *Journal of the Japanese Society of Agricultural Machinery, 61*(2), 61–69.
3. Albright, L. D., (2001). Controlling greenhouse environments. In: *International Symposium on Design and Environmental Control of Tropical and Subtropical Greenhouses* (Vol. 578, pp. 47–54).
4. Arima, S., Kondo, N., Shibano, Y., Fujiura, T., Yamashita, J., & Nakamura, H., (1994b). Studies on cucumber harvesting robot (part 2). *Journal of the Japanese Society of Agricultural Machinery, 56*(6), 69–76.
5. Arima, S., Kondo, N., Shibano, Y., Yamashita, J., Fujiura, T., & Akiyoshi, H., (1994a). Studies on cucumber harvesting robot (Part 1). *Journal of the Japanese Society of Agricultural Machinery, 56*(1), 55–64.
6. Artigas, J., Beltran, A., Jimenez, C., Baldi, A., Mas, R., Domınguez, C., & Alonso, J., (2001). Application of ion-sensitive field-effect transistor-based sensors to soil analysis. *Computers and Electronics in Agriculture, 31*(3), 281–293.
7. Åstrand, B., & Baerveldt, A. J., (2002). An agricultural mobile robot with vision-based perception for mechanical weed control. *Autonomous Robots, 13*(1), 21–35.
8. Auernhammer, H., Muhr, T., Searcy, S. W., Mitchel, B. W., Reid, J. F., & Ting, K. C., (1991). *GPS in a Basic Rule for Environment Protection in Agriculture* (Vol. 1, pp. 1–33). Wissenschaftszentrum Weihenstephan.
9. Baars, R. M. T., Solano, C., Baayen, M. T., & Rojas, J., (1996). MIS support for pasture and nutrition management of dairy farms in tropical countries. *Computers and Electronics in Agriculture, 15*(1), 27–39.
10. Bac, C. W., Van, H. E. J., Hemming, J., & Edan, Y., (2014). Harvesting robots for high-value crops: State-of-the-art review and challenges ahead. *Journal of Field Robotics, 31*(6), 888–911.
11. Bailey, B., (2004). Natural and mechanical greenhouse climate control. In: *International Symposium on Greenhouses, Environmental Controls,*

and In-house Mechanization for Crop Production in the Tropics 710 (pp. 43–54).

12. Barbari, M., (2005). *Planning Individual Showering Systems for Pregnant Sows in Dynamic Groups* (Vol. 7, p. 130). In: Livestock environment VII, 18–20 May 2005, Beijing, China American Society of Agricultural and Biological Engineers.
13. Bell, T., (2000). Automatic tractor guidance using carrier-phase differential GPS. *Computers and Electronics in Agriculture, 25*(1/2), 53–66.
14. Billingsley, J., & Schoenfisch, M., (1997). The successful development of a vision guidance system for agriculture. *Computers and Electronics in Agriculture, 16*(2), 147–163.
15. Blancheton, J. P., Piedrahita, R., Belaud, A., Moutounet, Y., Fievelstadt, S., & Lemarié, G., (2006). Gas control in land based aquaculture, physiological and technical aspects. *Aqua, 1*, 9–13.
16. Briskin, E., Kalinin, Y., Lepetukhin, K., & Maloletov, A., (2020). About Controlling the Shape of the Sprinkler. In: *Robotics: Industry 4.0 Issues and New Intelligent Control Paradigms* (Vol. 1, pp. 147–157). Springer, Cham.
17. Caffey, R. H., (1998). *Quantifying Sustainability in Aquaculture Production, 1*, 1–22.
18. Cao, Z. L., Huang, Y., & Hall, E. L., (1988). Region filling operations with random obstacle avoidance for mobile robots. *Journal of Robotic Systems*, *5*(2), 87–102.
19. Caponetto, R., Fortuna, L., Nunnari, G., Occhipinti, L., & Xibilia, M. G., (2000). Soft computing for greenhouse climate control. *IEEE Transactions on Fuzzy Systems*, *8*(6), 753–760.
20. Cardenas-Lailhacar, B., Dukes, M. D., & Miller, G. L., (2005). Sensor-based control of irrigation in bermudagrass. In: *2005 ASAE Annual Meeting* (Vol. 1, pp. 1–20). American Society of Agricultural and Biological Engineers.
21. Chang, C. M., Fang, W., Jao, R. C., Shyu, C. Z., & Liao, I. C., (2005). Development of an intelligent feeding controller for indoor intensive culturing of eel. *Aqua Cultural Engineering*, *32*(2), 343–353.
22. Chao, K., Chen, Y. R., Hruschka, W. R., & Park, B., (2001). Chicken heart disease characterization by multi-spectral imaging. *Applied Engineering in Agriculture*, *17*(1), 99.

23. Cho, S. I., & Lee, J. H., (2000). Autonomous speed sprayer using differential global positioning system, genetic algorithm, and fuzzy control. *Journal of Agricultural Engineering Research*, *76*(2), 111–119.
24. Choset, H., (2001). Coverage for robotics: a survey of recent results. *Annals of Mathematics and Artificial Intelligence*, *31*(1–4), 113–126.
25. Cohen, O., & Edan, Y., (2008). A sensor fusion framework for online sensor and algorithm selection. *Robotics and Autonomous Systems*, *56*(9), 762–776.
26. Costa, C., Loy, A., Cataudella, S., Davis, D., & Scardi, M., (2006). Extracting fish size using dual underwater cameras. *Aqua Cultural Engineering*, *35*(3), 218–227.
27. Csordas, A. T., Delwiche, M. J., & Barak, J., (2004). Automated real-time PCR biosensor for the detection of pathogens in produce irrigation water. In: *2004 ASAE Annual Meeting* (Vol. 1, pp. 1–20). American Society of Agricultural and Biological Engineers.
28. Cveticanin, D., & Wendl, G., (2004). Dynamic weighing of dairy cows: Using a lumped-parameter model of cow walk. *Computers and Electronics in Agriculture*, *44*(1), 63–69.
29. Daskalov, P. I., Arvanitis, K. G., Pasgianos, G. D., & Sigrimis, N. A., (2006). Non-linear adaptive temperature and humidity control in animal buildings. *Biosystems Engineering*, *93*(1), 1–24.
30. De Koning, K., (2004). Automatic milking, lessons from Europe. In: *2004 ASAE Annual Meeting* (Vol. 1, p. 1). American Society of Agricultural and Biological Engineers.
31. De Miranda, F. R., (2003). A site-specific irrigation control system. In: *2003 ASAE Annual Meeting* (Vol. 1, pp. 1–11). American Society of Agricultural and Biological Engineers.
32. De Villiers, T., (2008). *Fungal Enzymes and Microbial Systems for Industrial Processing* (Vol. 1, pp. 1–22). Doctoral Dissertation.
33. Delwiche, M., Tang, X., Bondurant, R., & Munro, C., (2001). Estrus detection with a progesterone biosensor. *Transactions of the ASAE*, *44*(6), 2003.
34. DeVries, T. J., Von, K. M. A. G., & Beauchemin, K. A., (2005). Frequency of feed delivery affects the behavior of lactating dairy cows. *Journal of Dairy Science*, *88*(10), 3553–3562.
35. Dohi, M., (1996). Development of multipurpose robot for vegetable production. *JARQ (Japan)*, 1, 1–22.

36. Edan, Y., & Miles, G. E., (1993). Design of an agricultural robot for harvesting melons. *Transactions of the ASAE, 36*(2), 593–603.
37. Edan, Y., Han, S., & Kondo, N., (2009a,b). Automation in agriculture. In: *Springer Handbook of Automation* (Vol. 1, pp. 1095–1128). Springer, Berlin, Heidelberg.
38. Eradus, W. J., & Jansen, M. B., (1999). Animal identification and monitoring. *Computers and Electronics in Agriculture*, *24*(1/2), 91–98.
39. Evett, S. R., Peters, R. T., & Howell, T. A., (2006). Controlling water use efficiency with irrigation automation: Cases from drip and center pivot irrigation of corn and soybean. In: *Proc. 28th Annual Southern Conservation Systems Conf.* (Vol. 28, pp. 57–66).
40. Farrell, J. A., Givargis, T. D., & Barth, M. J., (2000). Real-time differential carrier phase GPS-aided INS. *IEEE Transactions on Control Systems Technology*, *8*(4), 709–721.
41. Ferguson, H. G., (1925). *Apparatus for Coupling Agricultural Implements to Tractors and Automatically Regulating the Depth of Work* (pp. 1–44). Patent GB, 253566.
42. Figueiredo, G. F., Dawson, M. D., Benson, E. R., Van, W. G. L., & Gedamu, N., (2004). Advancement in whole house machine vision-based poultry behavior analysis. In: *2004 ASAE Annual Meeting* (Vol. 1, pp. 1–11). American Society of Agricultural and Biological Engineers.
43. Figueiredo, G. F., Dickerson, T. W., Benson, E. R., Van, W. G. L., & Gedamu, N., (2003). Development of machine vision-based poultry behavior analysis system. In: *2003 ASAE Annual Meeting* (Vol. 1, p. 1). American Society of Agricultural and Biological Engineers.
44. Filby, D. E., Turner, M. J. B., & Street, M. J., (1979). A walk-through weigher for dairy cows. *Journal of Agricultural Engineering Research*, *24*(1), 67–78.
45. Fisher, A. V., (1997). A review of the technique of estimating the composition of livestock using the velocity of ultrasound. *Computers and Electronics in Agriculture*, *17*(2), 217–231.
46. Frost, A. R., Schofield, C. P., Beaulah, S. A., Mottram, T. T., Lines, J. A., & Wathes, C. M., (1997). A review of livestock monitoring and the need for integrated systems. *Computers and Electronics in Agriculture*, *17*(2), 139–159.

47. Gavric, M., Martinov, M., Bojic, S., Djatkov, D., & Pavlovic, M., (2011). Short-and long-term dynamic accuracies determination of satellite-based positioning devices using a specially designed testing facility. *Computers and Electronics in Agriculture*, *76*(2), 297–305.
48. Gerrish, J. B., Stockman, G. C., Mann, L., & Hu, G., (1985). Image processing for path-finding in agricultural field operations. *Paper-American Society of Agricultural Engineers (USA). Microfiche collection. No. Fiche No. 85-3037*, pp. 1–44.
49. Gillis, K. P., & Giles, D. K., David, C. S., et al., (2001). *Injection and Fluid Handling System for Machine-Vision Controlled Spraying, 1*, 1–22.
50. Gordon, G. P., & Holmes, R. G., (1988). *Laser Positioning System for off-Road Vehicles* (Vol. 1, pp. 1–33). American Society of Agricultural Engineers (USA).
51. Grand d'Esnon, A., (1984). Robotic harvesting of apples. In: *Robotics and Intelligent Machines in Agriculture: Proceedings, First International Conference on Robotics and Intelligent Machines in Agriculture* (Vol. 1, pp. 1–33). Tampa, Florida. St. Joseph, Mich.: American Society of Agricultural Engineers.
52. Gray, S. A., (2002). *Planning and Re-planning Events for Autonomous Orchard Tractors*, (Vol. 1, pp. 1–40).
53. Grewal, M. S., Andrews, A. P., & Bartone, C. G., (2020). *Global Navigation Satellite Systems, Inertial Navigation, and Integration* (Vol. 1, pp. 1–33). John Wiley & Sons.
54. Grift, T., Zhang, Q., Kondo, N., & Ting, K. C., (2008). A review of automation and robotics for the bio-industry. *Journal of Biomechatronics Engineering*, *1*(1), 37–54.
55. Guo, L. S., Zhang, Q., & Han, S., (2002). Position estimate of off-road vehicles using a low-cost GPS and IMU. In: *2002 ASAE Annual Meeting* (Vol. 1, pp. 1–11). American Society of Agricultural and Biological Engineers.
56. Guo, L., Zhang, Q., & Han, S., (2002). Agricultural machinery safety alert system using ultrasonic sensors. *Journal of Agricultural Safety and Health*, *8*(4), 385.
57. Gygax, L., Neisen, G., & Bollhalder, H., (2007). Accuracy and validation of a radar-based automatic local position measurement

system for tracking dairy cows in free-stall barns. *Computers and Electronics in Agriculture*, *56*(1), 23–33.

58. Hadlow, M. L., (2016). *Wireless and Empire Ambition: Wireless Telegraphy/Telephony and Radio Broadcasting in the British Solomon Islands Protectorate, South-West Pacific (1914–1947): Political, Social, and Developmental Perspectives, 1*, 1–32.
59. Halachmi, I., Edan, Y., Maltz, E., Peiper, U. M., Moallem, U., & Brukental, I., (1998). A real-time control system for individual dairy cow food intake. *Computers and Electronics in Agriculture*, *20*(2), 131–144.
60. Haley, M. B., & Dukes, M. D., (2007). Evaluation of sensor based residential irrigation water application. In: *2007 ASAE Annual Meeting* (Vol. 1, pp. 1–19). American Society of Agricultural and Biological Engineers.
61. Han, S., Zhang, Q., Ni, B., & Reid, J. F., (2004). A guidance directrix approach to vision-based vehicle guidance systems. *Computers and Electronics in Agriculture*, *43*(3), 179–195.
62. Hannan, M. W., & Burks, T. F., (2004). Current developments in automated citrus harvesting. In: *2004 ASAE Annual Meeting* (Vol. 1, pp. 1–21). American Society of Agricultural and Biological Engineers.
63. Harrell, R. C., Adsit, P. D., Pool, T. A., & Hoffman, R., (1990). The Florida robotic grove-lab. *Transactions of the ASAE*, *33*(2), 391–399.
64. Hayashi, S., Ota, T., Kubota, K., Ganno, K., & Kondo, N., (2005). Robotic harvesting technology for fruit vegetables in protected horticultural production. *Information and Technology for Sustainable Fruit and Vegetable Production, 1*, 227–236.
65. Herwitz, S. R., Johnson, L. F., Dunagan, S. E., Higgins, R. G., Sullivan, D. V., Zheng, J., & Slye, R. E., (2004). Imaging from an unmanned aerial vehicle: Agricultural surveillance and decision support. *Computers and Electronics in Agriculture*, *44*(1), 49–61.
66. Hess, T., (1996). A microcomputer scheduling program for supplementary irrigation. *Computers and Electronics in Agriculture*, *15*(3), 233–243.
67. Holm, S., Brungot, J., Rønnekleiv, A., Hoff, L., Jahr, V., & Kjølerbakken, K. M., (2007). Acoustic passive integrated transponders for fish tagging and identification. *Aquacultural Engineering*, *36*(2), 122–126.

68. Hoyt, Jr. R. E., Hawkins, J. V., St Clair, M. B., & Kennett, M. J., (2007). Mouse physiology. In: *The Mouse in Biomedical Research* (Vol. 1, pp. 23-XVI). Academic Press.
69. Huang, Y., Cao, Z., & Hall, E., (1986). Region filling operations for mobile robot using computer graphics. In: *Proceedings 1986 IEEE International Conference on Robotics and Automation* (Vol. 3, pp. 1607–1614). IEEE.
70. Huhtala, A., Suhonen, K., Mäkelä, P., Hakojärvi, M., & Ahokas, J., (2007). Evaluation of instrumentation for cow positioning and tracking indoors. *Biosystems Engineering*, *96*(3), 399–405.
71. Iida, M., Furube, K., Namikawa, K., & Umeda, M., (1996). Development of watermelon harvesting gripper. *Journal of the Japanese Society of Agricultural Machinery*, *58*(3), 19–26.
72. Isidori, A., & De Persis, C., (2009). Feedback linearization of nonlinear systems. *Control Systems, Robotics and Automation-Volume XII: Nonlinear, Distributed, and Time Delay Systems-I*, *1*, 180–193.
73. Jaiswal, S., Benson, E. R., Bernard, J. C., & Van, W. G. L., (2005). Neural network modeling and sensitivity analysis of a mechanical poultry catching system. *Biosystems Engineering*, *92*(1), 59–68.
74. Jenkins, D. M., Delwiche, M. J., & Claycomb, R. W., (2002). Electrically controlled sampler for milk component sensors. *Applied Engineering in Agriculture*, *18*(3), 373.
75. Jin, J., & Tang, L., (2006). Optimal path planning for arable farming. In: *2006 ASAE Annual Meeting* (Vol. 1, pp. 1–20). American Society of Agricultural and Biological Engineers.
76. Jorgensen, R. N., Sorensen, C. G., Jensen, K., Sogaard, H. T., Pedersen, J. M., Sorensen, L. B., & Nielsen, J., (2006). Hortibot: An accessory kit transforming a slope mower into a robotic tool carrier for high-tech plant nursing-part I. In: *2006 ASAE Annual Meeting* (Vol. 1, pp. 1–14). American Society of Agricultural and Biological Engineers.
77. Juste, F., & Fornes, I., (1990). Contributions to robotic harvesting of citrus in Spain. In: *Proceedings of the AG-ENG 90 Conference, Berlin, Germany* (Vol. 1, pp. 146–147).
78. Juszka, H., Lis, S., & Tomasik, M., (2011). Validation of a model of the negative pressure set value signal formation in the cow machine milking. *Journal of Research and Applications in Agricultural Engineering*, *56*(2), 68–71.

79. Kassay, L., (1992). Hungarian robotic apple harvester. In: *American Society of Agricultural Engineers Meeting (USA)* (Vol. 1, pp. 1–30).
80. Katz, G., Arazi, A., Pinsky, N., Halachmi, I., Schmilovitz, Z., Aizinbud, E., & Maltz, E., (2007). Current and near term technologies for automated recording of animal data for precision dairy farming. *Journal of Animal Science*, *85*(suppl. 1), 377.
81. Kawamura, N., Namikawa, K., Fujiura, T., & Ura, M., (1984). Study on agricultural robot (part 1). *Journal of the Japanese Society of Agricultural Machinery*, *46*(3), 353–358.
82. Keicher, R., & Seufert, H., (2000). Automatic guidance for agricultural vehicles in Europe. *Computers and Electronics in Agriculture*, *25*(1, 2), 169–194.
83. KIM, S. J., & KIM, P. S., (2004). Optimal gate operation of irrigation reservoir using open water management program. In: *2004 ASAE Annual Meeting* (Vol. 1, p. 1–21). American Society of Agricultural and Biological Engineers.
84. Kim, Y., Evans, R. G., Iversen, W. M., Pierce, F. J., & Chavez, J. L., (2006). Software design for wireless in-field sensor-based irrigation management. In: *2006 ASAE Annual Meeting* (Vol. 1, pp. 1–11). American Society of Agricultural and Biological Engineers.
85. King, B. A., Wall, R. W., & Wall, L. R., (2000). Supervisory control and data acquisition system for closed-loop center pivot irrigation. *Supervisory Control and Data Acquisition System for Closed-Loop Center Pivot Irrigation. 1*, 1–8.
86. King, B. A., Wall, R. W., & Wall, L. R., (2005). Distributed control and data acquisition system for closed-loop site-specific irrigation management with center pivots. *Applied Engineering in Agriculture*, *21*(5), 871–878.
87. King, B. A., Wall, R. W., Kincaid, D. C., & Westermann, D. T., (2005). Field testing of a variable rate sprinkler and control system for site-specific water and nutrient application. *Applied Engineering in Agriculture*, *21*(5), 847–853.
88. Kise, M., Zhang, Q., & Noguchi, N., (2005). An obstacle identification algorithm for a laser range finder-based obstacle detector. *Transactions of the ASAE*, *48*(3), 1269–1278.
89. Kittas, C., Katsoulas, N., & Baille, A., (2001). Influence of greenhouse ventilation regime on the microclimate and energy partitioning of

a rose canopy during summer conditions. *Journal of Agricultural Engineering Research, 79*(3), 349–360.

90. Klocke, N. L., Hunter, C., & Alam, M., (2003). Application of a linear move sprinkler system for limited irrigation research. In: *2003 ASAE Annual Meeting* (Vol. 1, pp. 1–44). American Society of Agricultural and Biological Engineers.
91. Kondo, N., & Ting, K. C., (1998). Robotics for plant production. *Artificial Intelligence Review, 12*(1–3), 227–243.
92. Kondo, N., & Ting, K., (1998). *Robotics in Bio-Production Systems* (Vol. 1, pp. 1–22). St. Joseph: ASAE, The Society for engineering in agricultural, food, and biological systems.
93. Kondo, N., (1994). Control method for 7 DOF robot to harvest tomato. In: *Processing of the Asian Control Conference, 1994* (Vol. 1, pp. 1–4).
94. Kondo, N., (1995). Harvesting robot based on physical properties of grapevine. *Japan Agricultural Research Quarterly, 29*, 171–171.
95. Kondo, N., Monta, M., Shibano, Y., & Mohri, K., (1993a). Basic mechanism of robot adapted to physical properties of tomato plant. In: *Proceedings of the Korean Society for Agricultural Machinery Conference* (Vol. 1, pp. 840–849). Korean Society for Agricultural Machinery.
96. Kondo, N., Monta, M., Shibano, Y., & Mohri, K., (1993b). Two finger harvesting hand with absorptive pad based on physical properties of tomato. *Environment Control in Biology, 31*(2), 87–92.
97. Kondo, N., Monta, M., Shibano, Y., & Mohri, K., (1993c). Basic mechanism of robot adapted to physical properties of tomato plant. In: *Proceedings of the Korean Society for Agricultural Machinery Conference* (Vol. 1, pp. 840–849). Korean Society for Agricultural Machinery.
98. Kondo, N., Nishitsuji, Y., Ling, P. P., & Ting, K. C., (1996). Visual feedback guided robotic cherry tomato harvesting. *Transactions of the ASAE, 39*(6), 2331–2338.
99. Kurokami, K., (1980). Fence type training system of mandarin orange tree. *Agric. Hortic, 55*(2), 289–293.
100. Lam, Y., Slaughter, D. C., Wallender, W. W., & Upadhyaya, S. K., (2007). Machine vision monitoring for control of water advance in furrow irrigation. *Transactions of the ASABE, 50*(2), 371–378.
101. Lam, Y., Slaughter, D. C., Wallender, W. W., & Upadhyaya, S. K.,

(2006). Computer vision system for automatic control of precision furrow irrigation system. In: *2006 ASAE Annual Meeting* (Vol. 1, pp. 1–44). American Society of Agricultural and Biological Engineers.

102. Lee, P. G., (1995). A review of automated control systems for aquaculture and design criteria for their implementation. *Aquacultural Engineering, 14*(3), 205–227.
103. Levin, E., Zarnowski, A., Cohen, C. A., & Liimakka, R., (2010). Human centric approach to inhomogenious geospatial data fusion and actualization. In: *Proceedings of ASPRS* (Vol. 1, pp. 1–5).
104. Lines, J. A., Tillett, R. D., Ross, L. G., Chan, D., Hockaday, S., & McFarlane, N. J. B., (2001). An automatic image-based system for estimating the mass of free-swimming fish. *Computers and Electronics in Agriculture, 31*(2), 151–168.
105. Lopes, A. A., Paz, S. M., Cugnasca, C. E., & Saraiva, A. M., (2002). Sharing video images on the internet using Java: An application to controlled environment. In: *World Congress of Computers in Agriculture and Natural Resources, Proceedings of the 2002 Conference* (Vol. 1, p. 114). American Society of Agricultural and Biological Engineers.
106. Maltz, E., Devir, S., Kroll, O., Zur, B., Spahr, S. L., & Shanks, R. D., (1992). Comparative responses of lactating cows to total mixed rations or computerized individual concentrates feeding. *Journal of Dairy Science, 75*(6), 1588–1603.
107. Marchant, J. A., & Brivot, R., (1995). Real-time tracking of plant rows using a Hough transform. *Real-Time Imaging, 1*(5), 363–371.
108. Marchant, J. A., (1996). Tracking of row structure in three crops using image analysis. *Computers and Electronics in Agriculture, 15*(2), 161–179.
109. Martins, C. I. M., Eding, E. H., Verdegem, M. C., Heinsbroek, L. T., Schneider, O., Blancheton, J. P., & Verreth, J. A. J., (2010). New developments in re-circulating aquaculture systems in Europe: A perspective on environmental sustainability. *Aquacultural Engineering, 43*(3), 83–93.
110. Mast, M., Burmester, M., Krüger, K., Fatikow, S., Arbeiter, G., Graf, B., & Qiu, R., (2012). User-centered design of a dynamic-autonomy remote interaction concept for manipulation-capable robots to assist elderly people in the home. *Journal of Human-Robot Interaction, 1*(1), 96–118.

111. Matsuo, Y., Yamamoto, S., & Yukumoto, O., (2002). Development of tilling robot and operation software. In: *Automation Technology for Off-Road Equipment Proceedings of the 2002 Conference* (Vol. 1, pp. 180–184). American Society of Agricultural and Biological Engineers.
112. Merriam, J. L., Styles, S. W., & Freeman, B. J., (2007). Flexible irrigation systems: Concept, design, and application. *Journal of Irrigation and Drainage Engineering, 133*(1), 2–11.
113. Milla, K., & Kish, S., (2006). A low-cost microprocessor and infrared sensor system for automating water infiltration measurements. *Computers and Electronics in Agriculture, 53*(2), 122–129.
114. Miranda, F. R., Yoder, R. E., Wilkerson, J. B., & Odhiambo, L. O., (2005). An autonomous controller for site-specific management of fixed irrigation systems. *Computers and Electronics in Agriculture, 48*(3), 183–197.
115. Moltó, E., Pla, F., & Juste, F., (1992). Vision systems for the location of citrus fruit in a tree canopy. *Journal of Agricultural Engineering Research, 52*, 101–110.
116. Monta, M., (1997a). Basic study on chrysanthemum cutting sticking robot. In: *Proceedings of International Symposium on Agricultural Mechanization and Automation* (Vol. 1, pp. 93–98).
117. Monta, M., (1997b). Cutting providing system and vision algorithm for robotic chrysanthemum cutting sticking system. In: *Preprints of the International Workshop on Robotics and Automated Machinery for Bio-productions* (Vol. 1, pp. 7–12).
118. Morimoto, T., & Hashimoto, Y., (2000). An intelligent control for greenhouse automation, oriented by the concepts of SPA and SFA—an application to a post-harvest process. *Computers and Electronics in Agriculture, 29*(1, 2), 3–20.
119. Mozes, N., Zmora, O., Porter, C., & Gordin, H., (2003). Marine integrated pond system under desert conditions in southern Israel—potential, results and limitations. In: *Aquaculture Europe 2001 Conference* (Vol. 1, pp. 1–22). Trondheim, Norway.
120. Muir, J. F., Brugere, C., Young, J. A., & Stewart, J. A., (1999). The solution to pollution? The value and limitations of environmental economics in guiding aquaculture development. *Aquaculture Economics and Management, 3*(1), 43–57.
121. Nishiwaki, K., Amaha, K., & Otani, R., (2004). Development of nozzle

positioning system for precision sprayer. In: *Automation Technology for Off-Road Equipment Proceedings of the 2004 Conference* (Vol. 1, p. 74). American Society of Agricultural and Biological Engineers.

122. Njoroge, J. B., Ninomiya, K., Kondo, N., & Toita, H., (2002). Automated fruit grading system using image processing. In: *Proceedings of the 41st SICE Annual Conference. SICE 2002.* (Vol. 2, pp. 1346–1351). IEEE.
123. Noguchi, N., Ishii, K., & Terao, H., (1997). Development of an agricultural mobile robot using a geomagnetic direction sensor and image sensors. *Journal of Agricultural Engineering Research, 67*(1), 1–15.
124. Noguchi, N., Kise, M., Ishii, K., & Terao, H., (2002). Field automation using robot tractor. In: *Automation Technology for Off-Road Equipment Proceedings of the 2002 Conference* (Vol. 1, p. 239). American Society of Agricultural and Biological Engineers.
125. Norremark, M., Sorensen, C. G., & Jorgensen, R. N., (2006). HortiBot: Comparison of present and future phytotechnologies for weed control-part III. In: *2006 ASAE Annual Meeting* (p. 1). American Society of Agricultural and Biological Engineers.
126. O'Connor, M., Bell, T., Elkaim, G., & Parkinson, B., (1996). Automatic steering of farm vehicles using GPS. *Precision Agriculture, (precisionagricu3), 1*, 767–777.
127. Oksanen, T., Kosonen, S., & Visala, A., (2005). Path planning algorithm for field traffic. In: *ASAE Annual Meeting 2005* (Vol. 1, pp. 1–30).
128. Ollis, M., & Stentz, A., (1997). Vision-based perception for an automated harvester. In: *Proceedings of the 1997 IEEE/RSJ International Conference on Intelligent Robot and Systems. Innovative Robotics for Real-World Applications. IROS'97* (Vol. 3, pp. 1838–1844). IEEE.
129. Ordolff, D., (2001). Introduction of electronics into milking technology. *Computers and Electronics in Agriculture, 30*(1–3), 125–149.
130. Papandroulakis, N., Dimitris, P., & Pascal, D., (2002). An automated feeding system for intensive hatcheries. *Aquacultural Engineering, 26*(1), 13–26.
131. Park, J., & Nikravesh, P. E., (1996). A look-ahead driver model for autonomous cruising on highways. *SAE transactions, 1*, 1569–1577.
132. Pastell, M., Aisla, A. M., Hautala, M., Ahokas, J., Poikalainen, V., Praks, J., & Veermae, I., (2006). Automatic cow health measurement system in a milking robot. In: *2006 ASAE Annual Meeting* (Vol. 1, pp.

1–10). American Society of Agricultural and Biological Engineers.

133. Pastell, M., Takko, H., Gröhn, H., Hautala, M., Poikalainen, V., Praks, J., & Ahokas, J., (2006). Assessing cows' welfare: Weighing the cow in a milking robot. *Biosystems Engineering*, *93*(1), 81–87.

134. Pedersen, S. M., Fountas, S., Have, H., & Blackmore, B. S., (2006). Agricultural robots-system analysis and economic feasibility. *Precision Agriculture*, *7*(4), 295–308.

135. Peiper, U. M., Edan, Y., Devir, S., Barak, M., & Maltz, E., (1993). Automatic weighing of dairy cows. *Journal of Agricultural Engineering Research*, *56*(1), 13–24.

136. Perez-Munoz, F., Hoff, S. J., & Van, H. T., (1998). A quasi ad-libitum electronic feeding system for gestating sows in loose housing. *Computers and Electronics in Agriculture*, *19*(3), 277–288.

137. Peters, R. T., & Evett, S. R., (2005). Using low-cost GPS receivers for determining field position of mechanized irrigation systems. *Applied Engineering in Agriculture*, *21*(5), 841–845.

138. Pilarski, T., Happold, M., Pangels, H., Ollis, M., Fitzpatrick, K., & Stentz, A., (2002). The Demeter system for automated harvesting. *Autonomous Robots*, *13*(1), 9–20.

139. Poss, J. A., Russell, W. B., Shouse, P. J., Austin, R. S., Grattan, S. R., Grieve, C. M., & Zeng, L., (2004). A volumetric lysimeter system (VLS): An alternative to weighing lysimeters for plant-water relations studies. *Computers and Electronics in Agriculture*, *43*(1), 55–68.

140. Powell, J. M., Cusick, P. R., Misselbrook, T. H., & Holmes, B. J., (2007). Design and calibration of chambers for measuring ammonia emissions from tie-stall dairy barns. *Transactions of the ASABE*, *50*(3), 1045–1051.

141. Qiu, H., Zhang, Q., & Reid, J. F., (2001). Fuzzy control of electro hydraulic steering systems for agricultural vehicles. *Transactions of the ASAE*, *44*(6), 1397.

142. Rabatel, G., (1988). A vision system for Magali, the fruit picking robot. In: *Paper 88293, AGENG88, Int. Conf. Agricultural Engineering* (pp. 1–33). Paris.

143. Rabatel, G., Bourely, A., & Sevila, F., (1991). Object detection with machine vision in outdoor complex scenes: The case of robotic harvest of apples. In: *Engineering Systems with Intelligence* (Vol. 1, pp. 395–403). Springer, Dordrecht.

144. Rabatel, G., Bourely, A., Sevila, F., & Juste, F., (1995). Robotic harvesting of citrus: State-of-art and development of the French Spanish Eureka project. In: *Proc. International Conference on Harvest and Post-Harvest Technologies for Fresh Fruits and Vegetables* (Vol. 95, pp. 232–239).

145. Rajkondawar, P. G., Tasch, U., Lefcourt, A. M., Erez, B., Dyer, R. M., & Varner, M. A., (2002). A system for identifying lameness in dairy cattle. *Applied Engineering in Agriculture, 18*(1), 87.

146. Raut, M. J., & Shere, V. B., (2014). Automatic drip irrigation system using wireless sensor network and data mining algorithm. *International Journal of Electronics Communication and Computer Engineering, 5*(07), 195–198.

147. Reid, J. F., (2004). Mobile intelligent equipment for off-road environments. In: *Automation Technology for Off-Road Equipment Proceedings of the 2004 Conference* (Vol. 1, pp. 1–10). American Society of Agricultural and Biological Engineers.

148. Reid, J. F., Searcy, S. W., & Babowicz, R. J., (1985). *Determining a Guidance Directix in Row Crop Images* (Vol. 1, 1–44). American Society of Agricultural Engineers, USA.

149. Reid, J. F., Zhang, Q., Noguchi, N., & Dickson, M., (2000). Agricultural automatic guidance research in North America. *Computers and Electronics in Agriculture, 25*(1, 2), 155–167.

150. Ren, J., Buck, N. L., & Spahr, S. L., (1992). A dynamic weight logging system for dairy cows. *Transactions of the ASAE, 35*(2), 719–725.

151. Saeys, W., Mouazen, A. M., & Ramon, H., (2005). Potential for onsite and online analysis of pig manure using visible and near infrared reflectance spectroscopy. *Biosystems Engineering, 91*(4), 393–402.

152. Schofield, C. P., Whittemore, C. T., Green, D. M., & Pascual, M. D., (2002). The determination of beginning-and end-of-period live weights of growing pigs. *Journal of the Science of Food and Agriculture, 82*(14), 1672–1675.

153. Schueller, J. K., (1992). A review and integrating analysis of spatially-variable control of crop production. *Fertilizer Research, 33*(1), 1–34.

154. Sears, D. W. G., (2015). The explored asteroids: Science and exploration in the space age. *Space Science Reviews, 194*(1–4), 139–235.

155. Seipelt, F., Bunger, A., Heeren, R., Kähler, D., Lüllmann, M., & Pohl, G., (2003). Computer controlled calf rearing. In: *Fifth International*

Dairy Housing Conference for 2003 (Vol. 1, p. 356). American Society of Agricultural and Biological Engineers.

156. Sergeant, D., Boyle, R., & Forbes, M., (1998). Computer visual tracking of poultry. *Computers and Electronics in Agriculture*, *21*(1), 1–18.
157. Serodio, C., Cunha, J. B., Morais, R., Couto, C., & Monteiro, J., (2001). A networked platform for agricultural management systems. *Computers and Electronics in Agriculture*, *31*(1), 75–90.
158. Shani, U., (1980). Filling regions in binary raster images: A graph-theoretic approach. In: *Proceedings of the 7th Annual Conference on Computer Graphics and Interactive Techniques* (Vol. 1, pp. 321–327).
159. Shpigel, M., Neori, A., Popper, D. M., & Gordin, H., (1993). A proposed model for "environmentally clean" land-based culture of fish, bivalves, and seaweeds. *Aquaculture*, *117*(1, 2), 115–128.
160. Simonton, W., (1990). Automatic geranium stock processing in a robotic work cell. *Transactions of the ASAE*, *33*(6), 1274–1280.
161. Sittichareonchai, A., Sevila, F., Fatou, J. M., Constans, A., Brons, A., & Davenel, A., (1989). A robot to harvest grapes. In: *ASAE Paper, 89*, 7050–7074.
162. Slaughter, D. C., & Harrell, R. C., (1987). Color vision in robotic fruit harvesting. *Transactions of the ASAE*, *30*(4), 1144–1148.
163. Sorensen, C. G., Jorgensen, R. N., Pedersen, J. M., & Norremark, M., (2006). HortiBot: Application of quality function deployment (QFD) Method for horticultural robotic tool carrier design planning-Part II. In: *2006 ASAE Annual Meeting* (Vol. 1, pp. 1–33). American Society of Agricultural and Biological Engineers.
164. Steward, B. L., Tian, L. F., & Tang, L., (2002). Distance-based control system for machine vision-based selective spraying. *Transactions of the ASAE*, *45*(5), 1255.
165. Stombaugh, T. S., Benson, E. R., & Hummel, J. W., (1998). Automatic guidance of agricultural vehicles at high field speeds. *ASAE Paper, 1*, 983110.
166. Subrata, I. D. M., Fujiura, T., Nakao, S., Yamada, H., Hida, M., & Yukawa, T., (1997). 3-D vision sensor for cherry tomato harvesting robot. *Japan Agricultural Research Quarterly*, *31*, 257–264.
167. Tanigaki, K., Fujiura, T., Akase, A., & Imagawa, J., (2008). Cherry-harvesting robot. *Computers and Electronics in Agriculture*, *63*(1), 65–72.

168. Tasch, U., & Rajkondawar, P. G., (2004). The development of a soft separator™ for a lameness diagnostic system. *Computers and Electronics in Agriculture*, *44*(3), 239–245.

169. Teye, F., Gröhn, H., Pastell, M., Hautala, M., Pajumägi, A., Praks, J., & Ahokas, J., (2006). Microclimate and gas emissions in cold uninsulated dairy buildings. In: *2006 ASAE Annual Meeting* (Vol. 1, pp. 1–21). American Society of Agricultural and Biological Engineers.

170. Tillett, R. D., Frost, A. R., & Welch, S. K., (2002). Predicting sensor placement targets on pigs using image analysis. [Erratum: Oct 2002, v. 83 (2), p. 253.]. *Biosystems Engineering*, *81*(4), 453–463.

171. Ting, K. C., Giacomelli, G. A., & Fang, W., (1993). Decision support system for single truss tomato production. In: *Proceedings of XXV CIOSTA-CIGR V Congress* (Vol. 25, pp. 70–76).

172. Tomaszewski, M. A., Dijkhuizen, A. A., Huirne, R. B., & Otten, A., (2000). Management information system impact on dairy production for selected herds in Texas. *Computers and Electronics in Agriculture*, *26*(1), 1–11.

173. Torii, T., (2000). Research in autonomous agriculture vehicles in Japan. *Computers and Electronics in Agriculture*, *25*(1/2), 133–153.

174. Trevelyan, J. P., (1989). Sensing and control for sheep shearing robots. *IEEE Transactions on Robotics and Automation*, *5*(6), 716–727.

175. Turner, L. W., Anderson, M., Larson, B. T., & Udal, M. C., (2001). Global positioning systems (GPS) and grazing behavior in cattle. In: *Livestock Environment VI, Proceedings of the 6th International Symposium 2001* (Vol. 1, p. 640). American Society of Agricultural and Biological Engineers.

176. Upcraft, M. J., Noble, D. H., & Carr, M. K. V., (1989). A mixed linear programme for short-term irrigation scheduling. *Journal of the Operational Research Society*, *40*(10), 923–931.

177. Van Henten, E. J., (1994). *Greenhouse Climate Management: An Optimal Control Approach* (Vol. 1, pp. 1–22). Van Henten.

178. Van, A., M. A. P. M., Huirne, R. B. M., Dijkhuizen, A. A., Beulens, A. J. M., & Ten, C. A. U., (1999). Information needs and information technology on dairy farms. *Computers and Electronics in Agriculture*, *22*(2, 3), 97–107.

179. Wang, Y., Yang, W., Winter, P., & Walker, L. T., (2006). Non-contact sensing of hog weights by machine vision. *Applied Engineering in*

Agriculture, *22*(4), 577–582.

180. Wanjura, D. F., Maas, S. J., Winslow, J. C., & Upchurch, D. R., (2004). Scanned and spot measured canopy temperatures of cotton and corn. *Computers and Electronics in Agriculture*, *44*(1), 33–48.
181. Weiss, U., & Biber, P., (2011). Plant detection and mapping for agricultural robots using a 3D LIDAR sensor. *Robotics and Autonomous Systems*, *59*(5), 265–273.
182. Wheaton, F., & Hall, S., (2007). Research needs for automated oyster shucking. *Aquacultural Engineering*, *37*(1), 67–72.
183. Wiedemann, H. T., Ueckert, D. N., & McGinty, W. A., (2002). Spray boom for sensing and selectively spraying small mesquite on highway rights-of-way. *Applied Engineering in Agriculture*, *18*(6), 661.
184. Wilson, J. N., (2000). Guidance of agricultural vehicles—a historical perspective. *Computers and Electronics in Agriculture*, *25*(1/2), 3–9.
185. Wu, D., Zhang, Q., Reid, J. F., & Qiu, H., (1999). *Adaptive Control of Electrohydraulic Steering System for Wheel-Type Agricultural Tractors* (pp. 1–33). ASAE Paper, 993079.
186. Wurts, W. A., (2000). Sustainable aquaculture in the twenty-first century. *Reviews in Fisheries Science*, *8*(2), 141–150.
187. Xin, H., & Shao, B., (2005). *Real-time Behavior-Based Assessment and Control of Swine Thermal Comfort* (Vol. 7, p. 694). In: Livestock Environment VII, 18–20 May 2005, Beijing, China. American Society of Agricultural and Biological Engineers.
188. Yamashita, J., & Kondo, N., (1992). *Agricultural Robots (1): Vision Sensing Systems* (No. 92-3517, pp. 1–33). ASAE Paper.
189. Yitayew, M., Didan, K., & Reynolds, C., (1999). Microcomputer based low-head gravity-flow bubbler irrigation system design. *Computers and Electronics in Agriculture*, *22*(1), 29–39.
190. Zazueta, F. S., & Smajstrla, A. G., (1992). Microcomputer-based control of irrigation systems. *Applied Engineering in Agriculture*, *8*(5), 593–596.
191. Zhang, G., Strøm, J. S., Blanke, M., & Braithwaite, I., (2006). Spectral signatures of surface materials in pig buildings. *Biosystems Engineering*, *94*(4), 495–504.
192. Zhang, Q., & Qiu, H., (2004). A dynamic path search algorithm for tractor automatic navigation. *Transactions of the ASAE*, *47*(2), 639.

193. Zhang, Q., (1999). Hydraulic linear actuator velocity control using a feed forward-plus-PID control. *International Journal of Flexible Automation and Integrated Manufacturing*, *7*(3/4), 277–292.

194. Zion, B., Alchanatis, V., Ostrovsky, V., Barki, A., & Karplus, I., (2007). Real-time underwater sorting of edible fish species. *Computers and Electronics in Agriculture*, *56*(1), 34–45.

Chapter 3

Robotics for Agricultural Systems

CONTENTS

3.1. INTRODUCTION

In agriculture, robotics has been increasingly adopted, in the last few years, to improve efficiency and productivity. Most of the research efforts have been applied to fresh market vegetables and fruit harvesting jobs, which are usually tiring, time-consuming, and specifically demanding. Harvest labor for many crops accounts for about one-half to two-thirds of the overall labor costs. Additionally, due to a decrease in the population of farmers, harvesting is expected to be automated.

In applying robots to the variability of agricultural harvesting jobs, extensive research has been carried out: cucumbers, asparagus, tomatoes, apples, lettuce, citrus, oranges, grapes, strawberries, and strawberries. In (Arima et al., 2004; Brown, 2002; Edan et al., 2000; Hannan and Burks, 2004; Murakami et al., 1995; Peterson and Wolford, 2003b; Van Henten et al., 2002) some distinguished examples of agricultural robotic systems can be found. In recent years, Specific work on robotics for agricultural tasks such as spraying, harvesting, berry thinning and transplanting has been established (Ling et al., 2004; Monta et al., 1992; Monta et al., 1998). For developing visual guidance system, computer vision has also been extensively employed in agriculture (Benson et al., 2003; Pilarski et al., 1999), grade judgment of fruits (Nagata and Cao, 1998), for weed control (2002; Downey et al., 2003; Jeon et al., 2005) and or fruit recognition on trees (Peterson et al., 2003a). Specific research on vision-based harvesting of melons can be found in (Dobrusin et al., 1992), asparagus in (Humburg and Reid, 1992), and of tomatoes in (Chi and Ling, 2004) (Figure 3.1).

Figure 3.1: Lettuce harvest technology.

Source: http://www.gemuesetechnik.de/hortus-salaterntetechnik.html?&L=4.

3.2. RADICCHIO HARVESTER

Aiming at both cost-effectiveness and efficiency, the robotic harvester was designed (Foglia and Reina, 2006). It comprises of a grip and a double four-bar manipulator for the harvesting of radicchio. Radicchio needs a stem cutting of the plant about 10 mm under the soil surface, which is normally a 10–12 mm diameter stem and 120–130 mm in diameter. Both gripper and manipulators are pneumatically actuated. Pneumatic actuators have a high power-weight ratio though compared to electric actuators they are difficult to control, which makes them appropriate for agricultural applications. Moreover, the design of the gripper is according to the pneumatic muscles, which on the other hand are robust, light, easy to maintain, and inexpensive. Good compliance with the plant is also provided by the pneumatic actuation due to the compressibility of air, which permits to recompense for small errors in the measurement of radicchio location in the field (Kondo and Ting, 1998; Antonelli et al., 2017).

3.2.1. Grippers

The following requirements should be fulfilled by all grasping devices: simplicity in the mechanical design, robustness, easy implementation, and low-cost (Solis-Salazar et al., 2008; Zion et al., 2014).

A two-finger gripper prototype has been shown in Figure 3.2 designed for agricultural application; it employs two bucket-like fingers featuring a linear blade attached to their tips to carry out cutting operation and it is made of aluminum with a total weight of about 16 kg. Two pneumatic muscles actuate the driving linkage which is linked between the vertical slider *S* and the fixed plate *F* as specified in Figure 3.2(a, b); with symmetrical behavior, the two fingers operate simultaneously. Observe that the vertical stroke and of the slider S transforms into a vertical and horizontal displacement of the denoted fingertips with c and b in Figure 3.2(a). In the same figure, the dashed line shows the fingertip paths. When all four boundary switches touch the soil, the closure of the gripper starts. Then, at about 10 mm underground the fingers cut the stem and at the same time pull the plant out of the terrain as displayed in Figure 3.2(b) during the complete operation by the successive configurations of the gripper (Kelc et al., 2019).

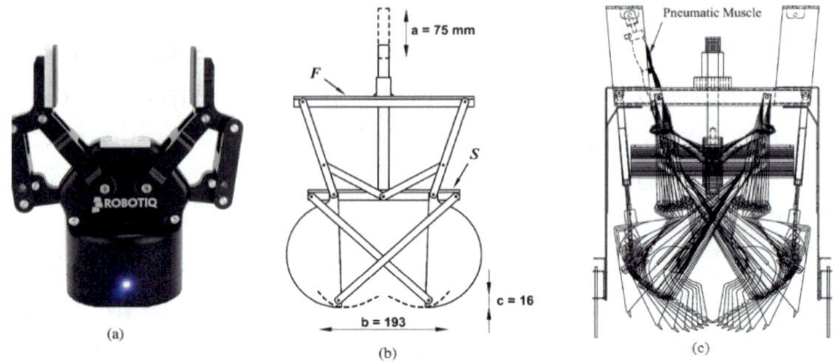

Figure 3.2*:* The two-finger gripper (a), and its cutting sequence (b).

Source: https://wiredworkers.io/product/robitiq-2f-85/.

Source: https://link.springer.com/chapter/10.1007%2F978-3-540-74027-8_27.

A three-finger gripper was also considered, to increase manipulation ability. It is shown in Figure 3.3 by the use of pneumatic muscles actuating three bucket-like fingers independently and a four-bar mechanism for driving, which permits performing the needed stem cutting at around 13 mm underground. This is displayed in Figure 3.3(b) by the series of cut trailed by one of the fingers. Note that the fingers are located at 120° and the height of the higher support for the muscles, signified with *B* in Figure 3.3(b) can be moved to regulate the relationship between the displacement of the fingertip and the pneumatic muscle stroke (Milella et al., 2004; 2006).

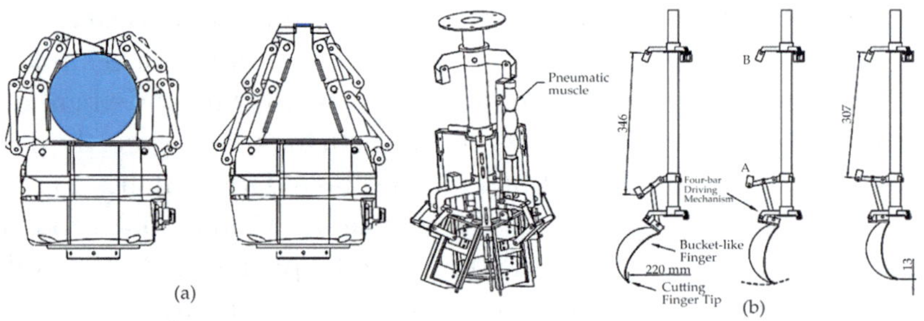

Figure 3.3: The three-finger gripper (a), and its cutting sequence (b).

Source: https://www.wevolver.com/wevolver.staff/3-finger.adaptive.robot.gripper/master/blob/Overview.wevolver.

3.2.2. Manipulator

To approach the plant, deliver the radicchio to a container on the carrier, and perform the harvesting task, the mobility to the gripper is provided by the manipulator.

Following are the requirements for the manipulator design: along the field lines, the displacement between the plants of about 700 mm, the speed of about 0.4 km/h of the carrier (tractor) of the robotic harvester, and the required minimum height of 800 mm from the ground by the CCD camera attached to the gripper in the field for efficient identification of the plant during the targeting stage. The basis of the architecture of the manipulator is on four-bar parallel links, which permit the gripper to stay level. Three applicant configurations are examined that use one, two, and three pneumatic actuators as shown in the functional schemes displayed in Figure 3.5. A moving delivery tray is used in all solutions reduce the harvesting cycle time. Normally, the number of independent degrees of freedom of the end-effector corresponds to the number of actuated degrees of motions of a manipulator. The more degrees of motions, the higher cost of the manipulator, and the higher the degree of flexibility (Sciavicco and Siciliano, 2000). Robot for sweet-pepper harvesting is shown in Figure 3.4.

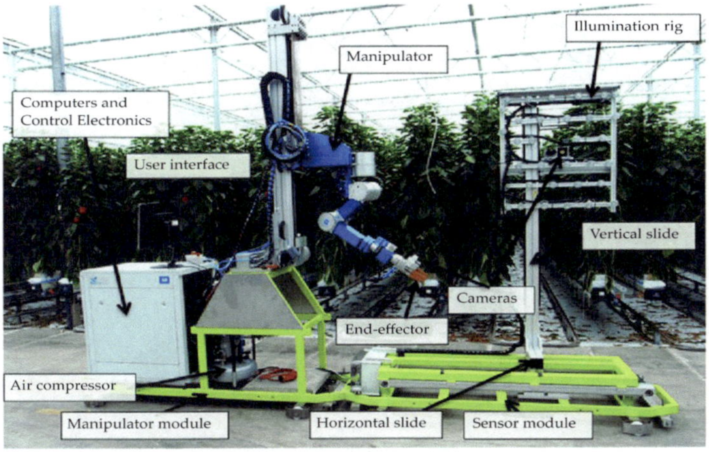

Figure 3.4: Robotic setup for sweet-pepper harvesting.

Source: https://www.researchgate.net/figure/Final-integrated-robot-for-harvesting-sweet-pepper-fruit-as-used-for-the-greenhouse_fig10_281536425.

The one-actuator architecture (Figure 3 5(a)) is less versatile but allows the lowest costs; hence, the best trade-off has been resulted by the two-actuator configuration (Figure 3.5(b)) and has been selected for this system. A characteristic harvesting cycle path has been shown in Figure 3.5 followed by the gripper with the two-actuator manipulator. To a ground reference frame in Figure 3.6(a), the same path is referred and to a coordinate system embedded with the carrier in Figure 3.6(b). Until the plant is localized by the vision-based module, the gripper moves forward horizontally (point *B*' in Figure 3.6(a, b)).

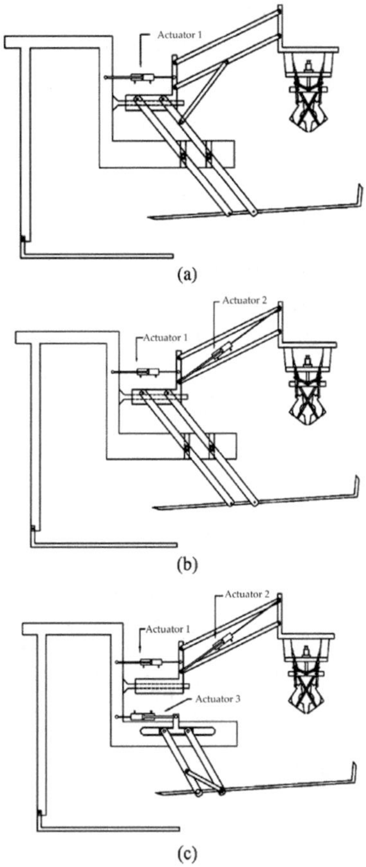

Figure 3.5: Four bar-based manipulator employing: one (a), two (b), and three actuators (c).

Source: https://link.springer.com/chapter/10.1007%2F978-3-540-74027-8_27.

Then, towards the plant (point *C*), the gripper starts its downward course, where it carries out the cutting operation and concerning the ground keeping null velocity. The return of the gripper towards the starting configuration has been marked by point *D* where the system can start the cycle again as the plant is dropped on the delivery tray (point *E*).

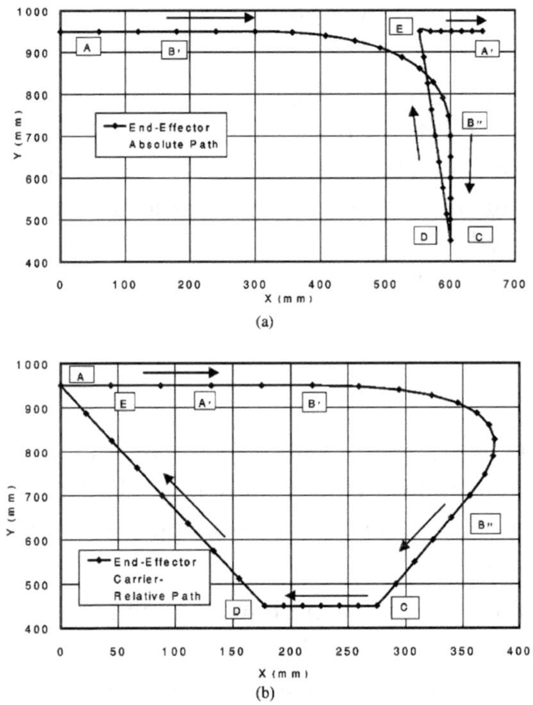

Figure 3.6: Harvesting cycle expressed in terms of the path of the end-effector concerning a ground frame (a), and to a carrier-embedded coordinate system (b).

Source: https://link.springer.com/chapter/10.1007%2F978-3-540-74027-8_27.

3.2.3. The Radicchio Visual Localization

To localize and detect the plants in the field, a vision-based algorithm was established; the basis of the RVL module morphological operations and intelligent color filters, which permit one to distinguish the radicchio within the images clutched by a CCD color camera with a frame rate of 5 Hz straddling on the end-effector.

Normally, the following steps are included in the algorithm:

- To enhance the thresholding operation, image acquisition in the Hue Saturation Luminance (HSL) space as described below (Figure 3.7(b)).
- To obtain two images, Hue, and Luminance plane extraction where the radicchio is physically separate from the surrounding (Figure 3.8).
- To obtain two binary images, independent thresholding in the Luminance and Hue planes comprising the radicchio white pixels and purple pixels respectively (Figure 3.9). By analyzing the histogram of the two planes, the thresholds are experimentally determined. Particularly, the threshold for the Hue plane was found to be well defined as:

$$\frac{T_{Max} + T_{Min}}{2} \quad (1)$$

where T_{min} and T_{max} are the minima and maximum intensity values respectively. The threshold for the Luminance plane is set, as the highest value of grey levels as the white parts of the plant related to the pixels with the largest luminance.

- To combine the information into a unique image, Or-operation of the two images (Figure 3.10(a)).
- Particle and morphological filtering (Figure 3.10(b)). Particularly, a morphological opening is applied using a 5×5 pixels substructure to remove isolated background pixels and open up touching futures, i.e., an operation of erosion followed by dilation. Then, objects with an area are filtered out, i.e., a total number of pixels, smaller than a threshold value (2000 pixels). Lastly, a morphological closing, by bridging the remaining small gaps allows reconstructing the shape of the plant, i.e., a succession of erosion and dilation.
- Convex hull production of the extracted feature (Figure 3.11(a)).
- Description of the minimum enclosing circle (Figure 3.11(a)), i.e., the minimum circle which encircles the extracted set of points (Xu et al., 2003). This geometrical algorithm assistances to notice the plant even when only a comparatively small uncentered part of the radicchio is extracted.

The overall result of the localization algorithm is shown in Figure 3.11(b). Over the original image, the minimum enclosing circle is overlaid with its center representing the coordinates of the centroid of the plant in the image reference frame.

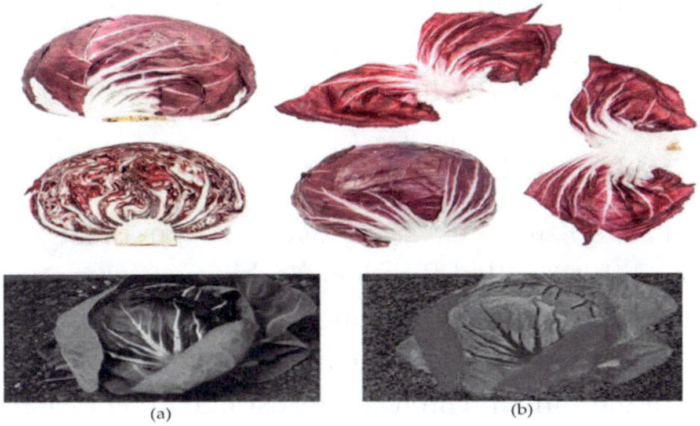

Figure 3.7: Sample image of the radicchio in the RGB (a), and HSL (b) color space.

Source: https://www.istockphoto.com/ae/photo/leaves-and-cabbageheads-of-red-radicchio-isolated-gm1172776579-325489497.

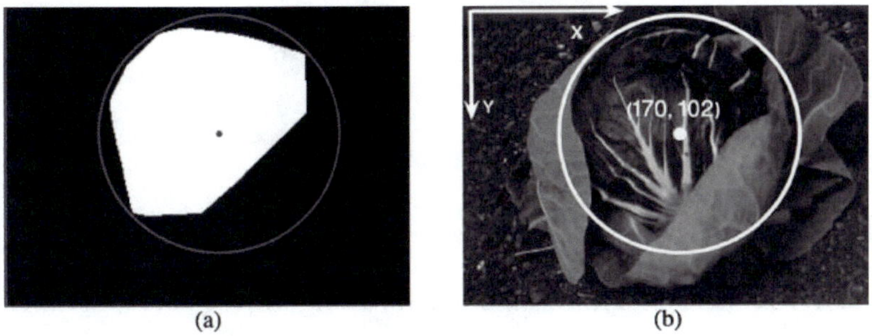

Figure 3.8: Convex hull and minimum enclosing circle generation (a), radicchio localization in the image plane (b).

Source: https://link.springer.com/chapter/10.1007%2F978-3-540-74027-8_27.

3.3. FENNEL CUTTING SYSTEM

A robotic system for the automated cut of just-harvest fennel wandering at high speed without any preordered alignment on a conveyor has been described in this section. During the post-harvest process of fennel, this operation is usually essential to generate ready-to-market products (Foglia et al., 2005; Ahmed et al., 2016). To enhance the quality of the final product and manufacturing rate, an automated cutting device would be helpful.

3.3.1. Cutting Mechanism

The proposed mechanism consists of two four-bar linkage cutters and is shown in Figure 3.9, so-called fore and backing cutter respectively, which function long the same path asynchronously that is shown by a black solid line in Figure 3.9. the time delay between 2 cutters is proportional to the required length of the funnel which can be estimated by visual identification module.

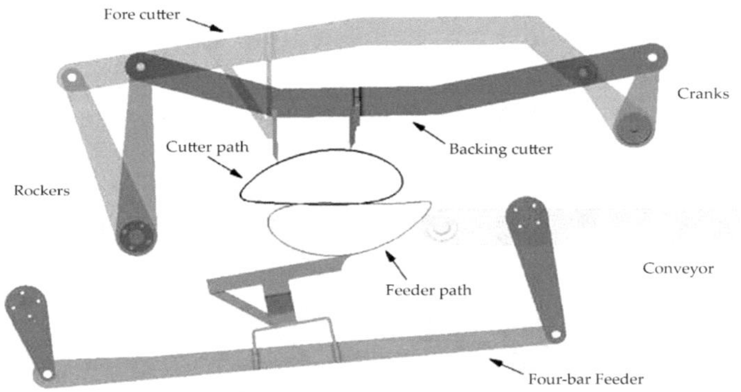

Figure 3.9: The mechanism for automated cutting of just-harvest fennel.

Source: https://link.springer.com/chapter/10.1007%2F978-3-540-74027-8_27.

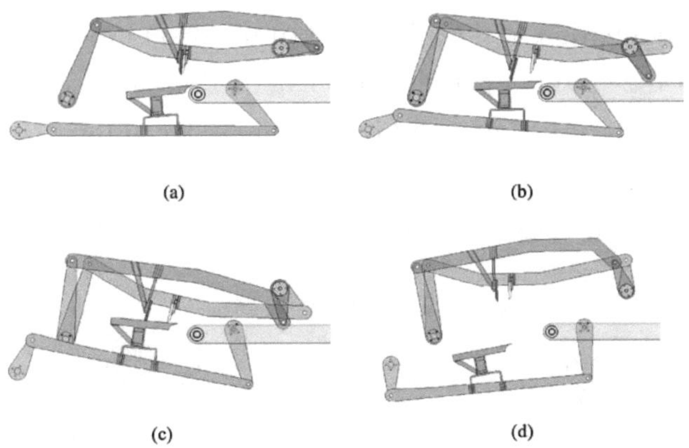

Figure 3.10: Configurations of the mechanism during a typical cutting cycle: starting configuration (a), first cut performed by the fore cutter (b), second cut performed by the backing cutter (c), and return of the two cutters to the starting configuration as the r.

Source: https://link.springer.com/chapter/10.1007%2F978-3-540-74027-8_27.

Proposed by the identification module defined later in this section. The plant from the conveyor is picked up by the third four-bar linkage feeder and delivers the needed support to carry out a double cutting operation. Observe that during the cutting stage the feeder and cutter follow about a straight line giving a strong and stable cutting action. Some typical configurations of the mechanism have been shown in Figure 3.8 during its working cycle (Anubhuti et al., 2011).

3.3.2. The Fennel Visual Identification

Fennels traveling at high speed on a conveyor are processed by the FVI system detecting the parts of the plant, such as leaves and root, and to produce a high-quality market product it needs to be cut off. In Figure 3 11 are shown, the final market product and the just-harvest fennel covered the cutting lines) delivered by the FVI module. The FVI module operates in real-time with a processing rate of 60–80 plants per minute and a sampling rate of 10 Hz. Usually, the following steps are included in the algorithm (Milella et al., 2006):

- RGB to HSL conversion and image acquisition.
- By processing the Hue plane, extraction of the plant from the image.

- Individuation of the alignment of the plant on the conveyor.
- For the cutting operation, the calculation of appropriate lines.

Each step is discussed in detail in the remainder of this section.

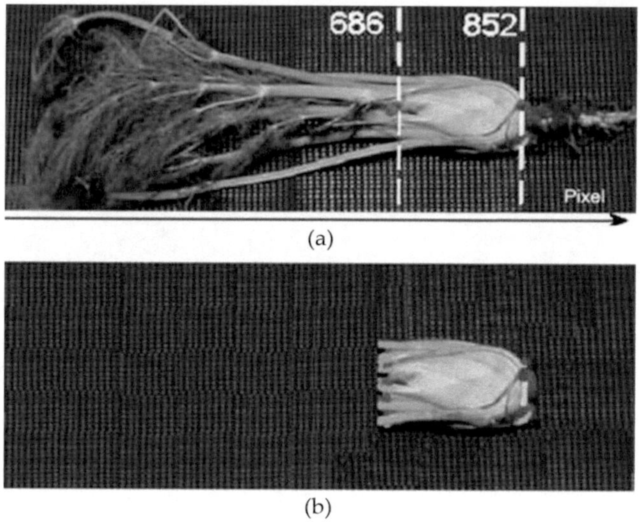

Figure 3.11: The FVI module establishes the cutting lines (a) to produce the ready-to-market fennel (b).

Source: https://link.springer.com/chapter/10.1007%2F978-3-540-74027-8_27.

- **Image Conversion in the HSL Space:** Instead is using RGB, the use of HSL space permits differentiating between luminance and color information, hence enhancing the successive image segmentation.
- **Hue Plane Processing:** Using thresholding in the Hue plane, the fennel can be effortlessly differentiated from the belt background by the related histogram. Through an initial calibration, the threshold value is experimentally determined. At the beginning of the process, an image of the belt is attained, without any fennel and the threshold for segmentation is fixed as μ-3σ where σ and μ are respectively, the standard deviation and the mean of the distribution of the grey levels of the hue image of the belt. The pixels of the plant can be designated and the grey levels of the belt are insulated. Then, joining connected pixels and eliminating isolated pixels can be performed by an appropriate combination

of dilation and erosion operations to gain a unique object demonstrating the fennel (Artés et al., 2002).

- **Individuation of the Plant Orientation and Cutting Lines:** it is essential to know the orientation of the plant, for the FVI to work correctly. Two approaches are adopted and are discussed below. The first approach aims to detect the relative position between the so-called "green" and "white" parts of the plant. By thresholding in the saturation plane, the *green* part can be detected, since it links to the pixels with the maximum intensity values in this plane.

Similarly, by thresholding in the luminance plane, the *white* part of the plant can be isolated, since this part is made up of pixels with the maximum intensity values in this plane. It is possible to derive the orientation of the plant by comparison of the position of the centroid of the green part with that of the white part. Then, considering both orientation and dimension of the plant, the cutting lines can be fixed. Particularly, either the right side or the left side of the bounding box, the rooted cutting line will be fixed according to the plant orientation, contouring the *white* part; at a distance from the root cutting line of two-thirds of the box length, the cutting line to removing the leaves is set (Escalona et al., 2005; Gao and Wang, 2011).

Based on color matching techniques, the second approach for approximation of the plant orientation is developed combined with the greyscale pattern matching. This method typically depends on the normalized cross-correlation approach. For the upper leaves and the root, two templates are defined respectively which will serve as masks. Each template is stimulated around the image approximating the value of the normalized correlation coefficient whose peak signifies the location of the best matches in the image of the mask (Masuhr and Menzel, 1972; Sacchetti et al., 2003).

3.4. EXPERIMENTAL APPLICATIONS OF AGRICULTURAL ROBOTICS

Field and experimental results are acquired by Foglia et al. (2008) are presented in this section to assess their performance in terms of repeatability, accuracy, variations in lighting and robustness to disturbances, obtained with our visual systems (Ruiz, 1999).

3.4.1. Radicchio Harvesting

Through laboratory tests, a viability study of the system was achieved along with field validation of the vision-based module. A prototype functioning in a test bed simulating quasi-real working conditions was used for the laboratory experiments. To optimize and set up components of the system, this stage was helpful. Then, in the field tests, the performance of the RVL module was authorized. The connection between the two four-bar linkages is provided by the three 3 mm thick steel plates as shown in Figure 3.12 consisting of 2 mm thick and 20 mm diameter aluminum tubes, between the carrier and the manipulator, and between the gripper and the manipulator. All the revolute joints are DryLin® bushings. At the end of the manipulator, a preliminary two-finger gripper was mounted. Without the gripper, the overall weight of the robotic arm is about 25 kg. Figure 3.12(b) shows the sparse fist-size rocks and the laboratory test bed set with characteristic agricultural terrain (Zheljazkov et al., 2013).

Figure 3.12: The robotic harvester prototype operating in the laboratory test bed.

Source: https://www.researchgate.net/figure/2-DOF-manipulator-prototype-test-bed-I_fig4_285801344.

3.4.2. Laboratory Tests

To assess the performance of the robotic harvester, a set of experiments was performed in picking up and identifying radicchios that were located randomly along the harvesting line. By the RVL module, the plant position was derived (Hatami et al., 2017).

The robustness of the algorithm to change in the lighting was also tested. The result of image segmentation is shown in Figure 3.13 under three dissimilar lighting circumstances attained by an adjustable video light. Even with a reduction of the environmental illumination level, the RVL works very well as much as 80% of the optimal value ($L = 0.8$), as shown in the bottom image of Figure 3.13.

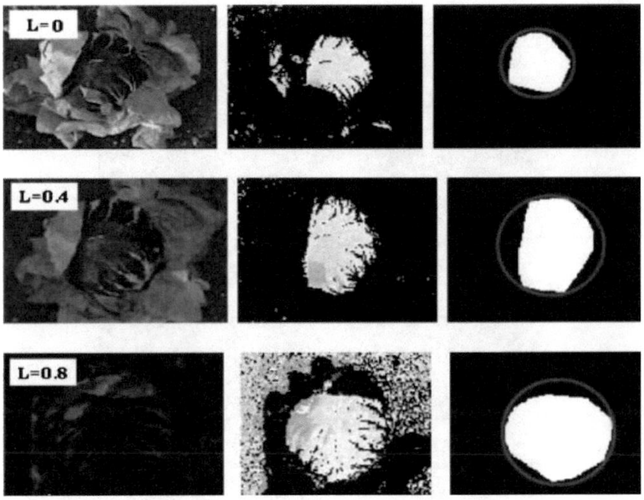

Figure 3.13: Lighting influence on the RVL module.

Source: https://link.springer.com/chapter/10.1007%2F978-3-540-74027-8_27.

3.4.3. Fennel Cutting

A laboratory test bed was set up, to test the performance of the FVI system using a ground-fixed firewire camera directing straight down 1 m across from a 2 m×0.8 m conveyor, simulating a characteristic post-harvest environment. To detect the leaf and root part of 30 cm lengthy fennel plants roaming on the conveyor with random alignment at the speediness of about 1.5 m/s, the FVI was extensively tested (Bell et al., 2008) (Figure 3.14).

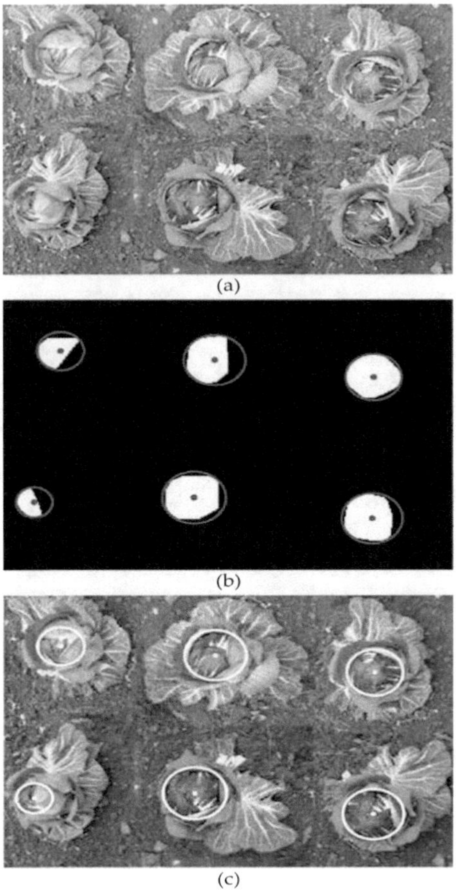

Figure 3.14: Plant detection in the field: acquired image (a), plant thresholding (b), and radicchio localization (c).

Source: https://link.springer.com/chapter/10.1007%2F978-3-540-74027-8_27.

The color filtering-based approach was adopted, in experiments of Foglia et al. (2006) to approximate the orientation of the plants. The lines of cutting for all the plants were set correctly by the system for a typical measurement as shown in Figure 3.15. In the experiments, no false detection was observed. The effect of lighting conditions on the performances on the FVI system was detected.

In Figure 3.16, the results are gathered showing good robustness and having a light reduction as much as 50% of the optimum value (L = 0.5). Though, for this application, lighting is predictable not to be a serious factor, as post-harvesting processes are usually achieved in structured environments.

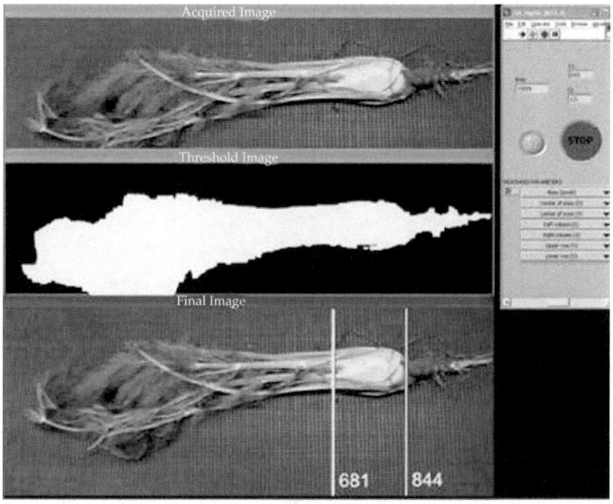

Figure 3.15: The typical measurement obtained from the FVI module.

Source: https://link.springer.com/chapter/10.1007%2F978-3-540-74027-8_27.

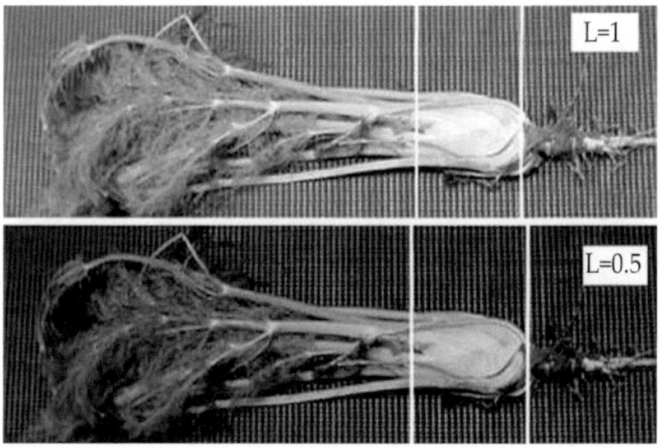

Figure 3.16*:* Lighting influence on the FVI module.

Source: https://link.springer.com/chapter/10.1007%2F978-3-540-74027-8_27

To estimate fennel orientation, implementing the template-based approach, experiments have also been accomplished; the cutting lines for this case are detected correctly by the FVI for only 80% of the plants. The differences in dimension and shape between plants are the cause of the lower performance, which is not accounted for in the template-based approach (Al-Dalain et al., 2012).

REFERENCES

1. Ahmed, M. F., Zayan, A. S., & Rashed, M. S., (2016). Evaluation of seed coating with certain bio-agents against damping-off and root rot diseases of fennel under organic farming system. *J. Phytopathol. Pst. Mgmt*, *3*, 11–23.
2. Al-Dalain, S. A., Abdel-Ghani, A. H., Al-Dala'een, J. A., & Thalaen, H. A., (2012). Effect of planting date and spacing on growth and yield of fennel (Foeniculum vulgare Mill.) under irrigated conditions. *Pak J Biol. Sci.*, *15*(23), 1126–32.
3. Antonelli, M. G., Zobel, P. B., Durante, F., & Raparelli, T., (2017). Development of an automated system for the selective harvesting of radicchio. *International Journal of Automation Technology*, *11*(3), 415–424.
4. Anubhuti, P., Rahul, S., & Kant, K. C., (2011). Standardization of fennel (Foeniculum vulgare), its oleoresin and marketed Ayurvedic dosage forms. *International Journal of Pharm. Sci. and Drug Res*, *3*(3), 265–269.
5. Arima, S., Kondo, N., & Monta, M., (2004). *Strawberry Harvesting Robot on Tabletop Culture* (pp. 1–30). ASAE Paper no. 04-3089. St. Joseph, Mich., ASAE.
6. Artés, F., Escalona, V. H., & Artés-Hdez, F., (2002). Modified atmosphere packaging of fennel. *Journal of Food Science*, *67*(4), 1550–1554.
7. Bell, C. E., Easley, T., & Goodman, K. R., (2008). Effective fennel (*Foeniculum vulgare*) control with herbicides in natural habitats in California. *Invasive Plant Science and Management*, *1*(1), 66–72.
8. Benson, E., Reid, J., & Zhang, Q., (2003). Machine vision-based guidance system for an agricultural small-grain harvester. *Trans. of the ASAE, 46*(4), 1255–1264.
9. Brown, G. K., (2002). *Mechanical Harvesting Systems for the Florida Citrus Juice Industry* (pp. 1–30). ASAE Paper no. 02-1108. St. Joseph, Mich., ASAE.
10. Chi, Y. T., & Ling, P., (2004). *Fast Fruit Identification for Robot Tomato Picker* (pp. 1–20). ASAE Paper no. 04-3083. St. Joseph, Mich., ASAE.
11. Dobrusin, Y., Edan, Y., Grinshpun, J., Peiper, U. M., & Hetzroni, A., (1992). *Real Time Image Processing for Robotic Melon Harvesting* (pp. 1–43). ASAE Paper No. 92-3515. St. Joseph, Mich., ASAE.

12. Doeney, D., Gilles, D. K., & Slaughter, D., (2003). *Ground Based Vision Identification for Weed Mapping Using DPGS* (pp. 5–23). ASAE Paper no. 03-1005. St. Joseph, Mich., ASAE.
13. Edan, T., Rogozin, D., Flash, T., & Miles, G. E., (2000). Robotic Melon Harvesting. *IEEE Trans. on Robotics and Automation, 16*(6), 831–834.
14. Escalona, V. H., Artés-Hernández, F., & Artés, F., (2005). Gas composition and temperature affect quality of fresh-cut fennel. *Hort. Science, 40*(3), 737–739.
15. Foglia, M. M., Gentile, A., & Reina, G., (2008). Robotics for agricultural systems. In: *Mechatronics and Machine Vision in Practice* (Vol. 1, pp. 313–332). Springer, Berlin, Heidelberg.
16. Foglia, M., & Reina, G., (2006). "Agricultural Robot for Radicchio Harvesting." *Journal of Field Robotics, 23*(6/7).
17. Foglia, M., Gentile, A., & Reina, G., (2005). A vision-based cutting system for fennel post-harvest processing. *Proc. of XVII AIMETA, Firenze, Italy, 1*, 11–15.
18. Gao, F. S., & Wang, M. Y., (2011). Breeding of a high yield and quality fennel variety' Denong no. 1'with cutting tolerance. *Northern Horticulture, 22*, 73.
19. Hannan, M. W., & Burks, T., (2004). *Current Developments in Automated Citrus Harvesting* (pp. 1–22). ASAE Paper no. 04-3087. St. Joseph, Mich., ASAE.
20. Hatami, T., Johner, J. C. F., & Meireles, M. A. A., (2017). Investigating the effects of grinding time and grinding load on content of terpenes in extract from fennel obtained by supercritical fluid extraction. *Industrial Crops and Products, 109*, 85–91.
21. Humburg, D. S., & Reid, J. F., (1992). Field performance of machine vision for the selective harvest of green asparagus. *Trans. ASAE, 100*(2), 81–92.
22. Jeon, H. Y., Tian, L. F., & Grift, T., (2005). *Development of an Individual Weed Treatment System Using a Robotic Arm* (pp. 1–32). ASAE Paper no. 05-1004. St. Joseph, Mich., ASAE.
23. Kelc, D., Vindiš, P., Rakun, J., Stajnko, D., & Lakota, M., (2019). Technology for a 'baby leaf' production of a corn salad and radicchio. In: *Proceedings of the 47th International Symposium, Actual Tasks on Agricultural Engineering* (Vol. 1, pp. 251–258). Opatija, Croatia. University of Zagreb, Faculty of Agriculture.

24. Kondo, N., & Ting, K., (1998). *Robotics for Bio-Production Systems* (Vol. 1, pp. 1–30). St. Joseph, Mi. ASAE.
25. Ling, P., Ehsani, R., Ting, K., Yu-Tseh, C., Ramalingam, N., Klingman, M., & Draper, C., (2004). *Sensing and End-Effector for a Robotic Tomato Harvester* (pp. 1–21). ASAE Paper no. 04-3088. St. Joseph, Mich., ASAE.
26. Masuhr, T., & Menzel, R., (1972). Learning experiments on the use of side—specific information in the olfactory and visual system in the honey bee (*Apis mellifica*). In: *Information Processing in the Visual Systems of Anthropods* (Vol. 1, pp. 315–321). Springer, Berlin, Heidelberg.
27. Milella, A., Reina, G., & Foglia, M., (2006). Computer vision technology for agricultural robotics. *Sensor Review*, *26*(4), 1–30, 290–300.
28. Milella, A., Reina, G., Foglia, M., & Gentile, A., (2004). Computer vision applications in agricultural robotics. In: *11th IEEE International Conference on Mechatronics and Machine Vision in Practice* (Vol. 11, pp. 1–33).
29. Monta, M., Kondo, N., & Ting, K. C., (1998). End-effector for tomato harvesting robot. *Artificial Intelligence Review, 12*, 11–25.
30. Monta, M., Kondo, N., Shibano, Y., Mohri, K., Yamashita, J., & Fujiura, T., (1992). *Agricultural Robots (3), Grape Berry Thinning Hand* (pp. 1–34). ASAE Paper No. 92-3519. St. Joseph, Mich., ASAE.
31. Murakami, N., Inoue, K., & Otsuka, K., (1995). Selective harvesting robot for cabbages. In: *Proc. of Int. Symposium of Automation and Robotics in Bio-Production and Processing* (Vol. 2, pp. 24–31.). JSAM,
32. Nagata, M., & Cao, Q., (1998). Study on grade judgment of fruit vegetables using machine vision. *Japan Agricultural Research Quarterly, 32*(4), 1–55.
33. Peterson, D. L., & Wolford, D., (2003b). Fresh-market quality tree fruit harvester part II, apples. *Applied Engineering in Agriculture, 19*(5), 545–548.
34. Peterson, D. L., Whiting, D., & Wolford, S. D., (2003a). Fresh-market quality tree fruit harvester part I, sweet cherry. *Applied Engineering in Agriculture, 19*(5), 539–543.
35. Pilarski, T., Happold, M., Pangels, H., Ollis, M., Fitzpatrick, K., & Stentz, A., (1999). The Demeter system for automated harvesting. In:

Proc. of the 8th Int. Topical Meeting on Robotics and Remote Systems (Vol. 8, pp. 10–44).

36. Ruiz, L., (1999). Fennel (*Faeniculum vulgare*) and native species. Examining a long term study on exotic species management methods and experiments on the basic establishment requirements of native species. *Senior Thesis* (Vol. 1, pp. 1–33). Board of Environmental Studies, University of California, Santa Cruz.
37. Sacchetti, G., Appendino, G., Ballero, M., Loy, C., & Poli, F., (2003). Vittae fluorescence as a tool to differentiate between poisonous and non-poisonous populations of giant fennel (Ferula communis) of the island Sardinia (Italy). *Biochemical Systematic and Ecology*, *31*(5), 527–534.
38. Sciavicco, L., & Siciliano, B., (2000). *Modeling and Control of Robot Manipulators* (Vol. 1, pp. 1–42). London, Springer-Verlag.
39. Solis-Salazar, M., González-Lutz, M. I., Mora-Acedo, D., & Duran-Quiros, A., (2008). Radicchio (Cichorium intybus) for export: Variety evaluation, size of seedlings for transplant and the use of a soil plastic cover. *Agronomía Mesoamericana*, *19*(2), 271–278.
40. Van, H. E. J., Hemming, J., Van, T. B. A. J., Kornet, J. G., Meuleman, J., Bontsema, J., & Van, O. E. A., (2002). An autonomous robot for harvesting cucumbers in greenhouses. *Autonomous Robots, 13*, 241–258.
41. Xu, S., Freund, R., & Sun, J., (2003). Solution Methodologies for the Smallest Enclosing Circle Problem. *Journal of Computational Optimization and Applications, 25*, 283–292.
42. Zheljazkov, V. D., Horgan, T., Astatkie, T., & Schlegel, V., (2013). Distillation time modifies essential oil yield, composition, and antioxidant capacity of fennel (*Foeniculum vulgare* Mill). *Journal of Oleo Science*, *62*(9), 665–672.
43. Zion, B., Mann, M., Levin, D., Shilo, A., Rubinstein, D., & Shmulevich, I., (2014). Harvest-order planning for a multiarm robotic harvester. *Computers and Electronics in Agriculture*, *103*, 75–81.

Chapter 4 Application of Machine Vision Systems in Crop Farming

CONTENTS

4.1. INTRODUCTION

Based on measuring, monitoring, and responding to variability in crops a farming management concept precision agriculture is formed. By optimizing the returns on input a decision support system for farm management is defined by the research on precision agriculture while preserving resources. By giving automated solutions to responsibilities that are conventionally, accomplished manually, precision agriculture has been supported widely by the use of machine vision. Manual approaches incline to be error-prone and tedious. To support agricultural activities, efficient, and accurate solutions can be provided by the machine vision. Additionally, the analysis of massive volumes of data can be achieved accurately and quickly by machine learning algorithms, giving away for the application of machine vision uses in agriculture.

Various views have been carried out focusing on the subject of machine vision due to modern developments in machine vision applications in agriculture. Pereira et al. (2017) reviewed the state-of-the-art published in the last decade for the automatic classification and detection of fruits about computer vision. Kamilaris and Prenafeta-Boldú (2018) reviewed 40 research articles that used deep learning techniques for applications in agriculture. Iqbal et al. (2018) presented a survey on detection of diseases in citrus plant leaves and their cataloging, while in the production of grains (maize, soybean, rice, barley, and wheat), a review on the application of computer vision is directed by Patrício and Rieder (2018). A general overview with technical support and detailed description is presented in Pajares et al. (2016) giving support regarding imaging sensors, spectral bands, optical systems, and geometric visual system arrangements regarding precise applications in agriculture. Reid et al., (2000) conducted a review to present the progress in research on automation of vehicles over the last 15 years in North America. The significant features of the latter review include vehicle motion models, navigation sensors, steering controllers, and navigation planners. Developments, research effort, and innovation in agricultural robots have been investigated and presented by Bechar and Vigneault (2016) for field operations and the limitations, gaps, and associated concepts. Recent innovations autonomous navigation systems have been focused on the review conducted in Shalal et al. (2013) in agricultural environments. A short overview of current status and worldwide development of precision agriculture technologies of recent years has been presented by research in Yaghoubi et al. (2013). Torii (2000) presents relevant research in autonomous agricultural vehicles in Japan. Ji et al. (2009) reviewed machine vision techniques used in agriculture. Machine vision technologies for image acquirement and their treating and investigation have been presented in this review in agricultural automation, with the main focus on two types of images, namely infrared and visible (Figure 4.1).

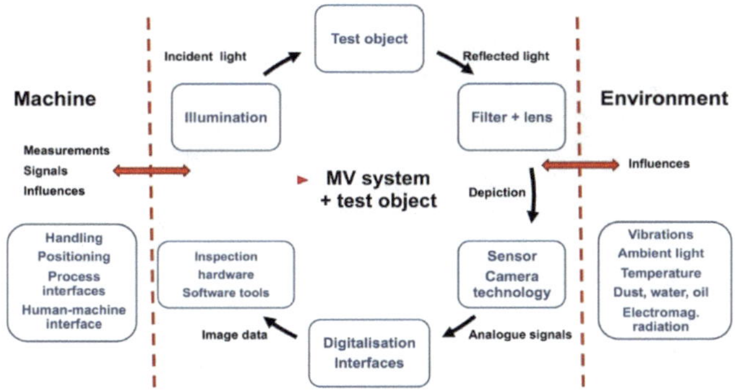

Figure 4.1: Schematic illustration of a machine vision system.

Source: https://www.stemmer-imaging.com/en-se/technical-tips/making-machine-vision-robust/.

Machine vision techniques in agriculture associated jobs are presented in this chapter focusing on the farming of crops. In the agricultural field, this work aims to provide a guide for practitioners and researchers delivering the recent developments on applications of machine vision in agriculture. Additionally, the review shown in this chapter not only addresses fruit type or plant (e.g., citrus) or precise agricultural challenge (e.g., disease detection) but also describes the research performed in various agricultural activities covering more technical topics, such as autonomous mobile agricultural robots and vision-based vehicle guidance systems (Figure 4.2).

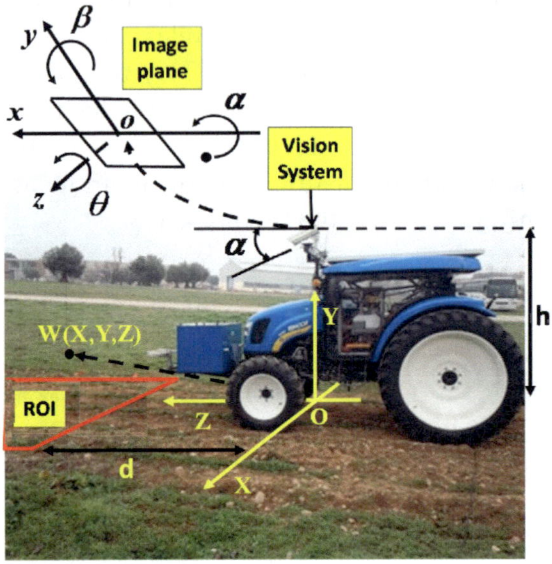

Figure 4.2: Application of machine vision in agriculture.

Source: https://www.mdpi.com/2313-433X/2/4/34/htm.

The following main matters are to be discussed in particular:

- Fruit and plant and detection methods;
- Harvesting support methodologies, including ripeness detection, fruit grading, yield prediction, and fruit counting;
- Fruit and plant disease detection and health protection methods, including weed, deficiency, disease, and insect detection;
- Types of the camera in agriculture used for machine vision;
- For agricultural applications, vision-based vehicle guidance systems (navigation); and
- Vision-based autonomous mobile agricultural robots.

It would be useful for researchers to be able to find all current operative machine vision methods applied for the broader classes of agricultural jobs, from the out-field progress of simple segmentation algorithms to automated vehicles through the in-field integration of refined algorithms, to study any automated precision agriculture engineering design. Hence, the above-mentioned six main issues have been addressed in this work regarding the above considerations, and before the suitable machine vision system is selected for installation for specific tasks in fields onboard an agricultural robot, these need to be considered. In precision farming applications, the proposed reviews on machine vision techniques toward the design of a (semi-)autonomous robot harvester on machine vision techniques as part of a national project entitled "Personalized Optimal Grape Harvest by Autonomous Robot" that is being recognized.

The compilation of this chapter is done according to the existing process of performing reviews by Kitchenham (2004). Articles reviewed were restricted to those available in English. The used search terms were "agricultural robots" and "machine vision in agriculture" which were keyed in the Google Scholar engine with a current publication date after 2017. Initially, the search resulted in 229 papers, but due to several reasons, some of them were discarded, e.g., not enough reference list, absence of a clear methodology, publishing houses of low reputation, etc. In the defined time-range for the machine vision-based robots in agriculture and corps, farming the use of machine vision in agriculture should be reported in the selected studies. This is the Benitti (2012) method. Though, rather than on the method of conducting the review, this work focuses on the results of the review.

4.2. PLANT AND FRUIT DETECTION APPROACHES

Separation of the plant from the rest of the plant and/or fruit from the background and the rest of the plant is achieved by image segmentation in agriculture. For carrying out agricultural activities automatically, this task is a key requirement like disease detection, yield estimation, and harvesting. Object detection can be done noninvasively, quickly, and accurately by image processing technologies, i.e., non-destructively for the fruits and plants. For several reasons, image processing for fruit and plant can be challenging (Badeka et al., 2019). Initially, the color degradation of the target's presence could be caused by various lighting conditions. Hence, illumination can seriously affect segmentation based on color information. When the target has a similar color to the background, the segmentation problem becomes even harder. For instance, it is problematic to distinct the fruit from the background when the target is a green fruit, like grape or apple containing branches and leaves which also have a green color. Segmentation based on color, in those cases, does not provide good results. Moreover, approaches that organize dissimilar kinds of features, such as shape and texture, and investigate a group of pixels and their connection to each other have been projected. Though, the usage of thresholds is required in many approaches for the features' (e.g., color, size, or shape) values and for any different image that needs to be defined, causing threshold-dependent performances. Image segmentation performance is the basis of the overall accuracy of plant and fruit detection. Hence, regardless of the surrounding environment and fruit color, a robust segmentation algorithm displaying high accuracy is needed. Therefore, a literature review of fresh research energies in fruit and plant detection is thought to be of great significance, due to the varying challenges. In what follows, appropriate fresh methods of the bibliography are presented.

In Behroozi-Khazaei and Maleki (2017) based on a genetic algorithm (GA) and an artificial neural network (ANN) presented a grape identification method using color features. To select superior color features and optimize the ANN structure simultaneously, a GA is employed. It does not have the same performance, although this technique performs well for a specific task, i.e., for mature grapes, when the color of the fruit is similar to the rest of the plant. This has the major benefit of segmentation algorithms that employ color features. Higher contrast is required between the background and target, to perform well. Li et al. (2018) presented a method that takes into account the above-mentioned disadvantage. In natural scenes, green apples can be detected by the proposed algorithm where the target has a similar color to the background. Initially, a close-loop graph represents the image with super pixels as nodes. Secondly, based on their similarity to the background, the nodes are ranked and foreground queries to produce the final saliency map. Lastly, to fit the V component in YUV color space Gaussian curve fitting is performed in salient areas, and to binarize the image a threshold is selected. Poor performance is displayed by this approach in segmenting the intense facula and edges with shadows around the object. Additionally, the performance of the technique is easily influenced by changing lighting conditions and is threshold-dependent.

To perform under uncontrolled and realistic conditions, an alternative method developed is proposed in Dias et al. (2018). To detect apple flowers, this technique is tested. On additional datasets, the generalization capability of the proposed approach is also examined. There are three stages in this method: (i) feature extraction based on Clarifai architecture a fine-tuned Convolutional Neural Network (CNN); (ii) classification of each region and to reduce the feature dimensionality, e application of Principal Component Analysis (PCA) according to the existence of flowers with a Support Vector Machine (SVM) classifier; (iii) computation of region proposals. In scenarios involving variations in illumination, color analysis alone leads to limited applicability which is among the research findings of this work. To detect cucumbers in a greenhouse Li et al. (2017) presented the technique based on both texture and color. For the identification task SVM is used. The state-of-the-art computer application systems are currently represented by deep learning techniques. However, a large amount of both computational power and data are required to fulfill the expectations for better performance. Additionally, the irregular growth of fruit and uneven illumination on the cucumber surface obstructed the segmentation task. Multiscale algorithms for multiple plant/fruit sizes are required to detect plants/fruits of various sizes. Based on texture, color, and shape, an image segmentation technique is presented in Prasetyo et al. (2017). This method applies Otsu thresholding on intensity, saturation, hue (HSV), and chromaticity blue, luminance, chromaticity red (YCbCr) color space for mango leaves. Segmentation results are capable, yet they are also unstable under varying illumination and are threshold-dependent.

In Sabzi et al. (2017), the aforementioned disadvantages, i.e., varying illuminance and various fruit/plant sizes are discussed. From morning to night, the intensity of sunlight changes, including a variation on color features. Hence, for all the conditions, the segmentation system needs to be proficient all day to be appropriate for real-time. Infield conditions, a new color space are suggested for segmenting green plants from their background. With diverse crop growth stages and under changing light intensities, the algorithm can achieve well. The recommended machine vision system contains two steps: the initial step contains the identification of the color condition and the other comprise of segmentation by the use of a hybrid artificial neural network-harmony search (ANNHS) classifier. For the classification of wheat grains to their species, a computer vision system using a multilayer perceptron (MLP) ANN projected in Sabanci et al. (2017). Using the Otsu method, images are converted to binarized, grayscale, and segmented using the thresholding operation. For each grain, the properties of color, texture, and size are captured, and the classification model is given as input. The main drawback of this technique is that images of grains are captured in artificial lighting (strip LED lighting) and not suitable for real environmental lighting conditions.

Yahata et al. (2017) presented a seedpod and a flower detection technique under completely outdoor environments in a crowd of soybean plants. Using image segmentation the detection of flower regions is carried out by simple linear iterative clustering (SLIC), which based on the distances between two pixels and the hue similarity, conducts pixel clustering (Achanta et al., 2012).

To make the final decision, CNN is introduced for flower detection. Features from an accelerated segment test (FAST) are used for flower counting to find saliency features (i.e., key points) for all the detected flower regions (Rosten et al., 2010) from the adjacent areas of the perceived flowers. Then, to find the matched points and for all the key points Oriented FAST and Rotated BRIEF (ORB) features are calculated (Rublee et al., 2011; Viola, 2004) to avoid counting the same flower twice in the two consecutive pictures due to overlapped regions. The Viola-Jones object detection technique is used for the detection of soybean seedpods. To improve the robustness against numerous lighting conditions Contrast limited adaptive histogram equalization (CLAHE) is carried out (Zuiderveld, 1994). Lastly, the candidate areas for seedpods are predicted by using CNN. Open issues remain although this technique returns good performances. Flowers of other plants positioned behind a target plant are included in these problems because they are detected together and leaves and stems sometimes detected wrongly as seedpods, and there are various cases where seedpods are mistakenly detected (Bishop, 2006; Duan et al., 2015; Lu et al., 2016).

In the field for rice panicles, an image segmentation method is implemented in Xiong (2017). Initially, to produce candidate regions SIFTSLIC is used. Secondly, as a candidate region classifier, a CNN (Lecun et al., 1998) is applied. Lastly, for optimizing the segmentation result, the entropy rate superpixel (ERS) algorithm is established. The proposed panicle segmentation (Panicle-SEG) algorithm shows higher segmentation accuracy as compared to other segmentation methods such as i2 hysteresis thresholding, hue plane threshold segmentation (HSeg), and joint segmentation (jointSeg) (Liu et al., 2011; Yang et al., 2012). Deep learning and parallel computing with the graphical processing unit (GPU) acceleration is accepted. However, the algorithm is yet not appropriate for real-time applications but runs relatively faster. Based on pulse coupled neural networks (PCNN) and an immune algorithm (IA), a cotton leaf segmentation method is presented in Zhang et al. (2018). By histogram statistical, three anti-light color mechanisms are selected with mean gray value. Then, the optimal number of iterations and the optimal parameters of the PCNN model is measured by the use of IA optimization so that the PCNN model can segment the objects in cotton regions more efficiently. Four image segmentation approaches, called the K-Means algorithm, the Otsu algorithm, PCNN, and FCM algorithm, are compared with the proposed technique to display that it has good resistance to complex background light changes.

Though the dependence on the iteration step is the disadvantage of this technique, and it has relatively poor real-time performance. Similar color weed as of cotton can also be detected as the target area. Based on the Sobel operator and Chan-Vese (C-V) model, an image segmentation technique of overlapping leaves is implemented in Wang et al. (2018). Initially, using a threshold the background is removed according to the relative levels of green in the image. C-V model is used for the extraction of the contour of the proposed leaf. Then, to perceive the edges of the leaf, eight directional Sobel operators are applied. Lastly, by combining the results the overlapping target leaf is extracted obtained by the Sobel operator and C-V model. The technique needs to be protracted to diverse kinds of crops, and in terms of robustness against environmental conditions, it is improved.

4.3. HARVEST SUPPORT APPROACHES

A critical agricultural activity harvesting is used for the gathering of fruit or ripe crops from the fields. For collecting and detecting the mature fruit/crops, intensive labor is needed. Based on numerous features like maturity level, size, damage level, and shape, a manual process of sorting is then carried out. However, manual sorting is subject to human errors and time-consuming, causing inconsistency in the quality of the final product. Hence, more efficient and accurate methods are required for assessing the crop or fruit collected. Compared to the manual approach, automated sorting of agriculture products is more efficient, which is fast, easy, and free from human errors. For supporting harvest, counting of fruits is another very significant task. In optimizing the harvest process, an automated and accurate method for counting of fruits can be helpful. To make cost-effective and more informed decisions, a better understanding of the inconsistency of yields across fields can be helpful for packaging, storage, labor allotment, and transportation (Oppenheim et al., 2017).

4.3.1. Fruit Grading and Ripeness Detection

Sorting of fruits is referred to as fruit grading based on parameters like shape, size, and maturity level. Due to human involvement in the sorting process, manual fruit grading leads to errors and is labor-intensive. Nowadays, for accurate and automated grading of fruits, machine vision is being used.

In Momin et al. (2017), a mango grading technique is proposed. Based on region global thresholding color binarization, an image processing algorithm combined with morphological analysis and the median filter is established to categorize mangos into one of three mass grades (small, medium, and large). The algorithm can be used in real-time applications and is very simple. Additionally, it can be applied to other crops as well because it is general. Though for experimenting fluorescent light may be effective, but algorithms that can perform well outside of the laboratory are needed for real-field applications. Wang et al. (2017) presented a method on mango size approximation. Histogram of oriented gradients (HOG) features combined in a cascade classifier is used. In the CIE Lab color space, Otsu's method is applied followed by color thresholding to eliminate background objects (branches, leaves, etc.). To remove the fruit pedicles, a 1D filter is developed, and to classify well-separated fruit an ellipse fitting technique is employed. Lastly, using the fruit image size, RGB-D depth information, and the thin lens formula, fruit lineal dimensions are calculated. This approach is appropriate for quick in-field estimation of fruit size, practical in terms of ease of use and cost, but cannot be applied directly under intense sunshine.

Utilizing image analysis, a method to approximate the minimum/maximum (equatorial/polar) mass and diameter length of olive fruits is presented by Ponce et al. (2018). Based on mathematical morphology, image segmentation is performed and the contrast between the background and the olives is maximized. Then, by automated thresholding, the olives are segmented based on the statistical bimodal analysis. The accuracy and robustness of the technique are underscored by the results. However, to confirm the generality of the technique to other cultivars, future trials in dissimilar lighting systems need to be examined. A technique for detecting the maturity (red, orange, and green) is presented by Wan et al. (2018) for fresh market tomatoes (Pear and Roma varieties). The feature color value in the proposed method is combined with the back propagation neural network (BPNN) classification method. Classification based on the color features as indicated by the results can be potentially used for the infield yield approximation of fresh tomatoes. Yet, identification results range within a substantial margin of error, the maturity of the tomatoes is green.

Abdulhamid et al. (2018) proposed a soya bean health and ripeness detection method based on texture and color features. The input images are transformed into HSI from the RGB color model. A gray-level co-occurrence matrix (GLCM) is used for extracting the statistical parameters such as homogeneity, contrast, energy, correlation, and texture features. Finally, based on the extracted features, an ANN is trained for the classification into been leaf, ripe, unhealthy, unripe, ripe but not ready for harvest. However, around the globe, the use of soya beans is diverse; the prediction of their ripeness periods will aid reduces losses. Under an uncontrolled environment, the used images are captured without familiarizing any disparity in brightness/intensity to the images, to gain their natural color.

4.3.2. Fruit Counting and Yield Prediction

For an approximation of the yield, the counting is very important. Using machine vision, the approximation of fruit count is quite challenging. Measuring the size and counting the numbers of fruits by machine vision is based on a recommendation that fruit is not obstructed by leaves and all fruit can be seen. To recognize each fruit in the tree image, with some fruit veiled within the canopy is the scientific challenge especially in the early period. Occlusion produced by the surrounding fruits and foliage and illumination can cause variation in the appearance of the fruit, leading to miscounting. However, for planning bins, labor, and harvest management as well as grading, storage, and transport. Recent research efforts are presented in this section which addresses the problem of fruit counting (Choi et al., 2017).

In Sethy et al. (2019a, b) presented using HSV color space for fitting with circular Hough transform (CHT) and for segmenting the flowers from the background for the counting and detection of marigold flowers. This approach has an advantage that it is capable of counting and detecting the marigold flower even in overlapping and/or occlusion conditions High-quality images and high-resolution cameras are needed for better results. A fruit counting and detecting approach based on CNN is implemented in Chen (2017). Based on a second convolutional network, a counting algorithm is then used for the approximation of the number of fruits, and mapping what fruit count approximates to a final fruit count a linear regression model is used. The issue of overlapping fruit is addressed in this approach which is an advantage of this method. However, due to the human-generated labels (ground-truth), it is prone to errors for the input data. A technique for citrus counting and detection is proposed in Dorj (2017). There are following steps involved in this method: thresholding, an adaptation of RGB image to HSV, noise removal, orange color detection, counting, and watershed segmentation. For the segmentation of objects, the watershed transform is effective with unclear boundaries. However, due to the existence of many local minima, it can cause over-segmentation. For that reason, marker-controlled watershed transformations are arranged. Fruits could be detected twice, which is the disadvantage of this method sue to their portion being split by the leaves of branches.

4.4. PLANT/FRUIT HEALTH PROTECTION AND DISEASE DETECTION APPROACHES

There are several factors which can affect the health of fruits and plants. Weeds are one of the most important factors which can influence the health of fruit and plant. Nutrition, moisture, and light are required by weeds which are shared with the growing plant or fruit. Hence, weed will have a negative influence on the quality and growth of the final product if it is not removed and detected early. For automating weed detection, machine vision has been used widely. When there is not a noteworthy difference between a weed and plant color, the job becomes very challenging. Insects are another factor that can have a negative influence on the health and growth of the plants. Automated monitoring of insects can help the farmers to protect the fruits and plants. For detecting deficiencies and diseases, machine vision has also been used. Regarding disease detection, there is a substantial amount of literature available. Mainly due to the variation on the texture and color of the diseased part, the detection of diseases is a challenging task. Disease detection methods can also be affected significantly by the variation in the lighting conditions (Cheng et al., 2017).

4.4.1. Weed Detection

The job of automatically detecting weeds is referred to as weed detection between crops. The quality of the final product and growth of the plant could be affected by weeds. Accurately distinguishing weeds from the plants is the main challenge. The presence of overlapping leaves, illumination variations, and insignificant color differences between crop and weed make the job more difficult.

Based on a CNN for the classification to carrots and weeds an HSV color space for image segmentation, a weed detection method is implemented in Knoll et al. (2018). This method based on color performs well due to the great difference of color between weeds and carrots. With unique and distinguishable colors, the color feature is most prominent. Therefore, it is possible that this technique could not precisely distinguish crop from weed, in cases where the difference in color in not significant. Moreover, variation in lighting conditions can affect the color-based methods.

For that reason, to segment, the weeds from the main crop, a texture-based approach based on ANNs and wavelet texture features is given by Bakhshipour et al. (2017). From numerous spatial orientations, wavelet transformation extracts information, which is useful for analyzing the image content, fractal analysis, and texture discrimination (Mallat, 1989). For converting the feature space to a lower number of dimensions, the principal component examination is used. It has been shown in the experimental results that even when there was a substantial amount of leaves overlapping and occlusion, the proposed method can discriminate crop from weed. Mainly in parts where the crop's leaf surface was, deformed or wrinkled, errors happened. However, errors were so small that they account for less than 1% of the misclassifications. For discriminating soil, crop, and weeds, the HSV color space is used. By filtering each of the HSV channels, the region of interest (ROI) is defined between definite values (maximum and minimum threshold values). Using a morphological dilation and erosion process, the region is then further refined, and to determine the mass distribution and the position of objects the moment method is applied to track crops in the video sequence. In the cataloging of various sorts of crop leaves, high accuracies can be obtained (Ali et al., 2017; Ebrahimi et al., 2017; Vassallo-Barco et al., 2017).

4.4.2. Insect Detection

For farmers, one of the main concerns is pest management. Marketable yields can be affected by crop damages and reduction in production loss. Hence, various methods are used by the farmers to protect and control the fields against damages of pests. During the last few years, the initial low cost, quick influence, the lack of knowledge on the part of farmers, easy accessibility of pesticides have increased their use causing serious consequences for animals, public health, and the environment. Hence, other than the use of chemical pesticides, better approaches to pest control are needed. Great damage can be caused by fruits and plants by insects. For instance, an invasive insect *Lobesia botrana* is thought to be one of the most harmful pests in vineyards (*Vitis vinifera L.*) (Chuanlei et al., 2017; Kounalakis et al., 2018). Expert knowledge is required for the conventional technique for the detection of insects and is also time-consuming. Moreover, for automatically identifying insects, machine vision has been deployed. Enhanced precision and work speed can be observed by the machine vision for automated insect identification also decreasing the human errors, since farmers may have partial pest scouting expertise (Bay et al., 2008; Lottes and Stachniss, 2017).

4.4.3. Disease and Deficiencies Detection

Automatic detection of the unhealthy parts of the plant is known as deficiency and disease detection. Due to the variability and importance of the subject, it is under research, a fact which makes the job difficult. The diseased part varies in texture and color depending on the type of plant of fruit, and type of deficiency and diseases. The appearance of the diseased part can also be affected by the variation in the lighting conditions (Milioto et al., 2018; Lottes et al., 2018).

Automatic identification of nutritional deficiencies of Iron (Fe), Boron (B), Potassium (K) and Calcium (Ca) has been focused in the study presented by Vassallo-Barco (2017), using descriptors of the texture and shape of the coffee leaves images. Using the Otsu method, the image segmentation is carried out, the GLCM and the blurred form model (BSM) descriptions to extract the texture and shape features. Then, by using the extracted features, neural network, naïve Bayes, k-Nearest Neighbors (k-NN) classifiers are trained to identify the kind of deficiency presented in each analyzed image. Though higher performance has been shown by the BSM descriptor and Bayes classifier than the rest of the methods, the general performance was shown to be very poor (F-measure 64.9%). Ali et al. (2017) presented a method for the detection of diseases in citrus plants. LBP (textural information) and HSV, RGB (color histogram) as descriptors, and Delta E (DE) are used for segmentation on the collected images. Cubic SVM, fine k-NN, bagged tree, and boosted tree ensemble classification approaches are used, and by using any color features compared to the other classifiers, the bagged tree ensemble classifier performs better. The experimental results showed that in the detection of plant diseases textural features, as well as the color features, are important. Further, in the detection of the diseased area, the combination of these features is also valuable in the plant. In Chuanlei et al. (2017) a correlation-based feature selection technique is proposed to find apple leaf disease. Shape, texture, and color features are applied as inputs to an SVM model to categorize apple leaf diseases. But the under natural illumination proposed technique fails to identify apple leaf diseases. In Zaw and Hlaing (2017), based on SIFT features a method for plant disease detection is proposed. To extract the whole region of the leaf the input image is preprocessed. From the pre-processed image the scale-invariant feature transform (SIFT) features are extracted. By a generalized Pareto distribution

(GP) the extracted SIFT features are modeled. To form the new proposed feature vector the two parameters of the GP model are along with GLCM numerical features. Finally, by using the extracted feature vectors an SVM classifier is proficient. The proposed features attain satisfactory classification accuracy. Nevertheless, although the prediction speed training time and of the classifiers may be suitable for mobile devices, but not suitable for real-time prediction (Salembier et al., 1998; Bosilj et al., 2018).

The C-V model is used in agricultural image processing, whereas previous information the data contents of a region of interest (ROI) are removed. In the C-V model nonetheless, the R, G, and B channels in the ROI are stationary, and for every channel, the color images and the weights are determined artificially (García et al., 2017; Pérez et al., 2017). In this chapter, PCA is used for picking three color channels from the G, H, B, R, S, and by calculating the contrast of the object pixel and background values adaptive weights of every channel are achieved. K-means grouping and used for unraveling the scrape region from the rest of the leaf (Chan and Vese, 2001). According to experimental results, this technique accomplishes improved segmentation precision than those of the initial Otsu segmentation with fewer iterations or an initial circle curve. However, when the studied regions are vague regarding the background, the proposed technique does not attain satisfying results (Otsu, 1979; Hamuda et al., 2017).

In Kumar and Patil (2017), a CBIR (content-based image retrieval system) for recovering diseased leaves of soybean is presented. The content-based image retrieval system (CBIR) based system contains two steps; feature matching and feature extraction. For this purpose, texture, shape, and color features are extracted. Precisely, by using HSV color histogram color features are extracted. In the form of matching main points, shape features are provided by the SIFT algorithm (Hu et al., 2017). Lastly, a new texture feature is proposed, i.e., local gray Gabor pattern (LGGP), by combining local binary pattern (LBP) and Gabor filter. When a query image is given the extracted feature vectors are stored and, its features are taken out, and to find the one with the smallest distance its distance to the feature vectors is calculated. The proposed system gives promising results. These outcomes are database dependent. If the size of the database is altered then the result may change. The leaf is segmented from the background with the determination of thresholds for b*, a*, and L* channels rather than using auto-threshold such as Otsu. Then, the GLCM is used for extracting statistical texture structures such as homogeneity, correlation, energy, and contrast. From the histograms of the extracted color planes, numerical indicators such as skew, standard deviation, energy, mean entropy, and are calculated. Lastly, on the extracted features to categorize the leaves into affected and healthy ones an SVM classifier is proficient. The classification rates of the technique are high, on a huge scale pathing the way to automated plant disease diagnosis. Possibly, more kinds of diseases affecting several types of plants will be combined with the system.

An image processing technique by using region growing and color information for segmenting greenhouse vegetable foliar disease spot images captured under real field situations is presented in Ma et al. (2017). When locally uneven lighting occurs the quality of color, index-based segmentation reduces intensely. For that purpose, a comprehensive color feature (CCF) is proposed which cartels ExR (excess red index) (Meyer et al., 1999), b* element of L*a*b* color space and H element of HSV color space, which are invariant to illumination (Zhou et al., 2014). Then, based on the CCF map a cooperative region rising method is used for disease spot separation from the clutter background. The experimental results show that for the segmentation of disease spots on images taken under real field circumstances the algorithm achieves better than K-means clustering and Otsu algorithm. Both the Otsu algorithm and the K-means clustering algorithm are affected by background noise and locally irregular illumination, producing a lot of false positives segmentations. In Prakash et al. (2017) a method for the classification and detection of diseases in citrus leaves is applied. With K-means clustering image segmentation is performed. GLCM surface features are extracted and used to drill an SVM classifier for the disease classification. Results are encouraging, and future work contains in different plant species the experimentation of diseases. In (Sharif et al., 2018), for citrus fruit and leaves a disease classification and detection method are presented. First, to improve the contrast of input images by applying a Gaussian function and top-hat filter image pre-processing is performed. Precisely, the top-hat filter is applied to increase the visual quality of the input image, which is by a difference-Gaussian image further improved. A weighted segmentation method is proposed for the segmentation of the disease graze spots, which uses threshold and chi-square distance function. Then, in a codebook geometric, texture, and color features are fused. By a hybrid feature selection method feature selection is done, which comprises entropy, PCA score, and skewness-based covariance vector. The particular features are specified as inputs to a multi-class support vector machine (M-SVM) for citrus disease classification. A complete comparison displays that the proposed method outdoes the current methods in the literature when tested on similar datasets. Future work comprises a deep model to be useful to the particular citrus datasets, as deep learning is described to perform well in the field of computer vision. Nevertheless, to be implemented the latter requires a big dataset.

4.5. CAMERA TYPES USED FOR MACHINE VISION IN AGRICULTURE

The choice of the suitable camera equipment for a precise machine vision application in crop farming establishes a fairly laborious task, considering the diverse cameras that are accessible in the market. Thus, some selection measures need to be defined by the designer considering mostly the application necessities that need to be satisfied. This section aims to give valuable visions to researchers by outlining the common kinds of equipment used for obtaining images/videos for machine vision applications in agriculture according to the literature. The frame rate, camera's sensor resolution, image connectivity (transfer rate), and worth seem the most significant factors that necessary to be studied to select a suitable camera that fits the needs of each application (Zhang et al., 2017; 2018).

According to the application, the camera device can be obvious based on three wider categories, namely: stereovision cameras, multispectral cameras, and RGB cameras. The most affordable and simple vision system involves a single RGB camera. This is the key reason why they are widely used in agricultural machine vision applications for segmentation tasks, yield prediction, fruit/plant detection, weed detection, ripeness detection, disease detection, and insect detection. As a substitute reasonable and convenient solution for image taking, the RGB build-in cameras of mobile devices are also used. At what time using RGB cameras, the potential for operative image processing is because of the use of effective image analysis algorithms, instead of image quality. For image display, RGB color space is satisfactory, but not for color processing, because the intensity is not decoupled from the chromaticity and there is a high connection among the components B. F and R, G or this reason, for algorithms that use color processing RGB images are not suitable. Furthermore, it is well-known that the color info in an image is very significant in image segmentation, but it is very subtle to lighting. The harmful effects of non-uniform illumination by normalizing the color vectors may be decreased, but they are not eliminated (Zhang and Sha, 2013; Garcia-Lamont et al., 2018). On top of that, the complex background increases any induced difficulty. The quality of images relies on the camera's basic properties dealing with distance from the object, sizes of objects, and lighting conditions. Cameras with higher resolution are favored for better results.

Humans observe colors in the three-wide channels B, G, and R, but by higher-precision color, measurement plants are potentially distinguished (McCarthy et al., 2010). Some studies use ultrasonic camera technologies, referring to hyper-spectral, thermal, or multispectral images for objects (leaves, plants, insects, fruit, etc.) detection. These technologies usually provide improved results than common RGB color images. This is for the reason that objects with the same color may display different reflectance in non-visible regions, and thus, they are simply distinguished with non-conventional RGB cameras. Conversely, compared with RGB cameras these technologies need expensive devices, and so for practical use are not affordable. The combination of the two methods, i.e., the adding of multispectral filters to a low-cost image system, has the potential to give features of attention strongly associated with biased wavelengths (Liu et al., 2017; Radcliffe et al., 2018).

Moreover, the variation of distinct plants, like weed and crops, in a 2D image is difficult. As an extra parameter, the dimension of crops (leaf area, crop height, leaf shape, depth information, etc.) can be observed. With stereo vision, this can be achieved. 3D plant structure resulting from stereoscopic cameras is used in applications to species discrimination and plant and crop monitoring.

The above studies disclose that for image acquisition there are various types of cameras available, depending on the precise agricultural application. The critical feature in all cases is the camera's resolution, however which should be the adequate height to capture the particulars of the scene particularly in the cases of disease and insect detection. Furthermore, when there are desires for symbolizing machine vision abilities to an autonomous vehicle (robot), the practice of an industrial camera with high-speed communication capabilities (e.g., GigE) and improved frame rate is, in some sense, predictable.

4.6. VISION-BASED VEHICLE GUIDANCE SYSTEMS FOR AGRICULTURAL APPLICATIONS

For crop farming after studying the most typical methods in machine vision applications and the camera equipment applied to implement them, it would be very useful to review the vision-based vehicle systems planned in the literature that deployed these types of machine vision methods.

For a longer duration of work, automated agricultural vehicles are very useful, as an independent vehicle may outlive a human worker, increase operation safety, increase application precision, and productivity. Autonomously driven robots can aid farmers to decrease the manpower and hours required to achieve the desirable agricultural tasks, and at the same time as navigation is automatic conduct other high-level tasks. Furthermore, compared to common methods autonomous navigation is economically and environmentally friendly, of achieving the same task, because vehicles do not walk around the fields worthless but navigate optimally to exact areas of attention by selecting the shortest paths. Therefore, in terms of money and time (i.e., salaries, and fuel) farmers can well use their resources, thus saving money and for the masses to economize an improved product. Autonomous vehicles no longer are at the fantasy level with these paybacks in mind but are becoming practicable, and they are desirable nowadays. Automatic vehicle navigation in agricultural applications is taken as a separate field of science for these reasons.

Sophisticated sensors are being used by early guiding systems, representing technical feasibility (Warner and Harries, 1972). Low reliability and high cost were the key reasons for the non-commercialization of these systems. Today, navigation is sturdily linked with positioning information of the vehicle in a local or a global coordinate system. Global positioning systems (GPS) are generally used as global supervision sensors. The low accuracy of GPS receivers is the chief disadvantage of GPS-based navigation. Common GPS precision is around ±50 cm, which in the case of precision farming is an important tolerance (Yukumoto et al., 2000).

To improve the precision of navigation machine vision is used complementary to sensors and GPS. In a field when crop row structure is separate, in the field machine vision can be applied to automatically guide a vehicle. On one hand, the camera (guiding sensor) is only local, because regarding the relative location of the vehicle it discloses information. Alternatively, the use of local features can be used to fine-tune the vehicle navigation course (Bell, 2000; Han et al., 2004). To simulate the eyes of a human operator machine vision offers technological characteristics, thus, navigation systems have great potential (Wilson, 2000). The challenge of navigation in plant structures is the statistic that mowing or harvesting changes the section of the lines of the field. However, in most cases, in rows the plants grow, a natural set of lines is being produced, which describes the image, so making the processing more discrete. In common, autonomous navigation systems for agricultural mobile robot contains navigation control strategies, computational methods, and navigation sensors (Figure 4.3).

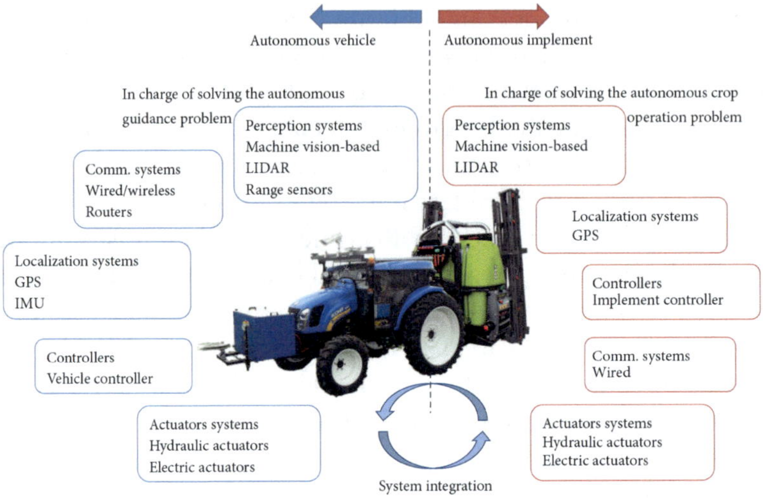

Figure 4.3: Novel trends in robotics for agricultural use.

Source: https://www.hindawi.com/journals/tswj/2014/404059/.

To define the directrix for a vehicle image processing method has been applied to crop row images. In what follows, the most topical methods of the literature are briefly described (Marchant, 1996).

The platform planned in Zhang et al. (1999) uses a feed forward-plus-PID steering controller and an E/H steering system in favor of automated guidance actuation. Terminated posture sensor, containing geomagnetic direction sensors (FOG), GPS, and for giving automated guidance information machine vision sensing systems are used on this stage. A strong procedure with row division by K-means gathering algorithm, row detection by a moment algorithm, and by a cost function guidance line selection is proposed in Han et al. (2004). For the development of the guidance directrix secondary information, like identified crop row spacing, is used. To guide an autonomous vehicle via the field in Bakker et al. (2008, 2011) based on grey-scale Hough transform on logically merged images a vision-based row detection system is used. In a sugar beet field in a greenhouse the experiment takes place. A technique to distinguish crop rows to guide agricultural robots to function in real-time even under an extensive range of illumination situations as described in Baker et al. (2008). To estimation, the position of the crop rows for image strips image pre-processing is led to get the binarization image and vertical projection technique. By Hough transform, the detection of crop rows is defined. There are three steps in the machine vision-based autonomous navigation system in Zhao and Jiang (2010). Firstly, to get the relationship among the image coordinates and the world coordinates the camera calibration is useful. Secondly, image processing and pattern recognition are used to get a quasi-navigation baseline. Finally, through Hough transform the real navigation line is removed from the quasi navigation baseline (Figure 4.4).

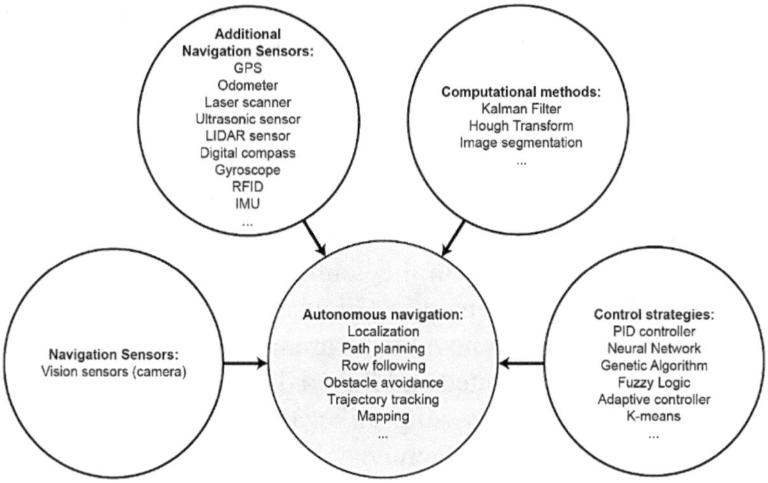

Figure 4.4: Basic parts of an autonomous vision-based navigation system.

Source: https://www.mdpi.com/2313-433X/5/12/89.

An orchard's color image by a multilayer feedforward neural network is classified into orchard elements in Torres-Sospedra and Nebot (2011). By Hough transform after a filtering operation the extraction of the favorite path is performed. A novel vision-based method for navigation of agricultural mobile robots in orchards is proposed in Chen and Sharifi (2015). By the mean-shift algorithm, the captured color image is gathered. A classification technique then based on graph dividing theory categorizes clustered image into defined classes containing sky, trees, and terrain. To extract the essential features to define the required central path for robot navigation in orchard rows Hough transform is applied. A real-time kinematic differential global positioning system (RTK-DGPS)-based independent field navigation system including automated promontory turns is presented in Baker et al. (2011) to offer a technique for crop row mapping combining machine vision, and to estimate the paybacks of a behavior-based reactive layer in hybrid thoughtful system architecture. Vision-based row guidance is offered in Xu and Xue (2010). To detect and identify crop plants the camera is used, and then constant and accurate navigation information is taking out from the binary image, which comprises the balance and heading angle of the robot comparative to the guidance row from the image. The vehicle in Scarfe et al. (2009) practices a mixture of GPS and machine vision to circumnavigate kiwifruit orchards, maneuver around obstacles like posts, and recognize braces. To guide a vehicle in a row of trees in Hiremath et al. (2014), a machine vision sensor is planned. The basic idea of using a machine vision for navigation is for computing the comparative position of the vehicle about a landmark and use that to approximate the vehicle's heading. The aim of Hiremath et al. (2014) is to advance an algorithm to separately guide a small mechanical ground vehicle platform with an orchard row, following the path of the row by using an upward-looking camera collective with a controller based on feature recognition from the contrast among the sky and the tree canopies. The technique presented in Hiremath et al. (2014) is based on a particle filter (PF) by using a novel measurement model, from the particle was a model image is constructed and linked directly with the measurement image after elementary processing, like thresholding, excessive-green filtering, and downsampling. Machine vision and later (laser radar) are independently used for guidance in Subramanian et al. (2006). To provide feedback on the steering angle a rotary encoder is used. To lessen the path error a PID controller is developed. Inflexible test paths constructed of mutual hay bales the vehicle's guidance precision is tested. In Wang et al. (2011) a stereo image processing algorithm is established which tracks and detects ground features taken in two consecutive field images, attained by using a vehicle-mounted stereo camera. To compute the lateral offset these ground features are used as reference points. The exact measurement of the vehicle's side offset can aid to get navigation accuracy. A flexible field-of-view machine vision method is established in Xue et al. (2012) that in cornfields allowing an agricultural robot to circumnavigate between rows. Using an image processing algorithm guidance lines are noticed, engaging morphological features in a near, far, and lateral field of view, and along these lines using fuzzy logic control the robot was guided. In the technique presented in Jiang et al. (2013), natural frames are examined in RGB vector space to research the possibility of curve path detection of shapeless roads. Morphological image processing and perceptual color clustering have been used as pre-processing to get the segmented path. From the right to left the pixels of the segmented image are examined, line-by-line to find the middle points of the path. To get the optimal navigation path, the least-squares curve fitting technique is applied, and finally, a new technique of finding the navigation parameters is presented. To process images in Meng et al. (2015), the HSI (hue, saturation, intensity) color model is used, and a verge algorithm based on the H component is used to yield grayscale images. According to the features of the crop rows in the image, a technique of crop line identification based on in lines scanning is proposed. To control the agricultural implement fuzzy control is used.

4.7. VISION-BASED AGRO-BOTS

A new problem is being introduced by agricultural practices, namely the essential to implement guidance technology, for example, a row weeder or a fertilizing machine along a crop row, with extreme precision. This was typically done manually by a person having a direct opinion of the rows. With the current growing concept, this is becoming impossible, which is in terms of time and money demanding. For agricultural practices, compared to the wide implementations the narrow row spacing is proposed, would need very quick reactions and careful handling by the human operator. Furthermore, in massive fields, the industrialized application would be overwhelming for a human operator. Automatically directed agricultural vehicles can decrease an operator's work intensity, and will not fatigue, which results in increased operational safety and improved efficiency. Therefore, the need for developing systems for automatic guidance regarding crop rows has become authoritative. Robots the agricultural robots were introduced to spread on accuracy techniques for agricultural operations like spraying, weeding, harvesting, transportation, etc. Such methods are also applied in the application of chemicals that are positioned in such a manner as to have an ideal effect with the least quantity or placement of fertilizer nearly the crop to lessen the source of nutrients to the weeds. The later also includes the new but growing concern of farmers about the environmental effects of agriculture, not just regarding severe chemicals, but also homicide water. To sense targets online and to work on a plant scale with extreme precision is the main goal of robots. Agro-Bots can give the farmer with up-to-date and exact information to assist management decisions, and save labor costs, safe workers from performing dangerous operations. Therefore, evolving Agro Bots is one of the tasks that agriculture is facing in developed countries, particularly those derived from ever-increasing production costs and labor deficiency (Søgaard and Lund, 2007; He et al., 2011).

Although industrial robots are extensively used for automation, it is somewhat rare to meet robots in agriculture. Relatively to agriculture fields, the industrial environment is, predictable, dry, clean, and well-lit while fields give random conditions in terms of terrain, weather, and light. Furthermore, industrial automation includes uniform, precise components robust for robotic operation, whereas agricultural automation deals with crops which differ in several ways; position, size shape, color, are often covered by foliage and are complex to handle. In structured environments Agro-Bots have been introduced (i.e., a simulated plantation in laboratories), along with different indoor (i.e., greenhouses) and outside agricultural environments. Though, only a few visions guided systems have been effectively tested and developed in real field trials (Figure 4.5).

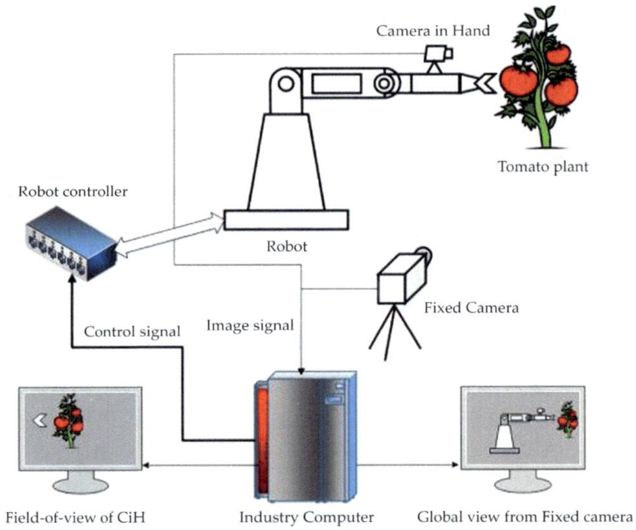

Figure 4.5: Vision-based agro-bots.

Source:https://www.sciencedirect.com/science/article/abs/pii/S0168169916304227.

In two main categories, the study on Argo-Bots is mostly divided. On tractors and other agricultural machinery, the first category is based to advance intelligent systems with navigation functions as described in the prior section. The other category is to individualistically develop intelligent mobile stages for agriculture, as described in the present section. But the last intelligent platforms are used to solve precise problems on automation of agricultural production, and they are restricted to definite functions. In other words, they refer to precise application environments, therefore, it is not easy to improve and expand. Furthermore, the commercialization of Agro-Bots is rare because of their partial periodic usage, which indirectly raises the cost of agricultural production. The consistency of a machine to achieve exact farming work rather than a human is another problem that keeps Agro-Bots off the market. The intellect level of Agro-Bots desires to be enhanced more to gain the farmer's assurance. Though for an Agro-Bot to achieve higher production high intelligence is essential, it is very significant to be: (1) reasonable, in terms of what it offers, (2) simple in its structure, (3) easy to operate, and (4) should be adaptive to the needs of the agricultural production. A simple structure refers also to a slight size. A robot of a lesser size is more easily movable and perhaps less expensive. It should be examined if crop production may be made cheaper, better, quicker, and with a group of small robots instead of a few large ones. Furthermore, one of the advantages of smaller robots is that they may become suitable and common to the non-agricultural community, for local personal use instead of mass custom (Muscato et al., 2005; Ortiz and Olivares, 2006).

REFERENCES

1. Abdulhamid, U., Aminu, M., & Daniel, S., (2018). Detection of soya beans ripeness using image processing techniques and artificial neural network. *Asian J. Phys. Chem. Sci., 5*, 1–9.
2. Achanta, R., Shaji, A., Smith, K., Lucchi, A., Fua, P., & Süsstrunk, S., (2012). SLIC super pixels compared to state-of the-art super pixel methods. *IEEE Trans. Pattern Anal. Mach. Intell., 34*, 2274–2282.
3. Ali, H., Lali, M. I., Nawaz, M. Z., Sharif, M., & Saleem, B. A., (2017). Symptom based automated detection of citrus diseases using color histogram and textural descriptors. *Comput. Electron. Agric., 138*, 92–104.
4. Badeka, E., Kalabokas, T., Tziridis, K., Nicolaou, A., Vrochidou, E., Mavridou, E., Papakostas, G. A., & Pachidis, T., (2019). Grapes visual segmentation for harvesting robots using local texture descriptors. In: *Computer Vision Systems (ICVS 2019)* (Vol. 1, pp. 98–109). Springer: Thessaloniki, Greece.
5. Bakhshipour, A., Jafari, A., Nassiri, S. M., & Zare, D., (2017). Weed segmentation using texture features extracted from wavelet sub-images. *Biosyst. Eng., 157*, 1–12.
6. Bakker, T., Van, A. K., Bontsema, J., Müller, J., & Van, S. G., (2011). Autonomous navigation using a robot platform in a sugar beet field. *Biosyst. Eng., 109*, 357–368.
7. Bakker, T., Wouters, H., Van, A K., Bontsema, J., Tang, L., Müller, J., & Van, S. G., (2008). A vision-based row detection system for sugar beet. *Comput. Electron. Agric., 60*, 87–95.
8. Bay, H., Ess, A., Tuytelaars, T., Van Gool, L., (2008). Speeded-Up Robust Features (SURF). *Comput. Vis. Image Underst., 110*, 346–359.
9. Bechar, A., & Vigneault, C., (2016). Agricultural robots for field operations: Concepts and components. *Biosyst. Eng., 149*, 94–111.
10. Behroozi-Khazaei, N., & Maleki, M. R., (2017). A robust algorithm based on color features for grape cluster segmentation. *Comput. Electron. Agric., 142*, 41–49.
11. Bell, T., (2000). Automatic tractor guidance using carrier-phase differential GPS. *Comput. Electron. Agric., 25*, 53–66.
12. Benitti, F. B. V., (2012). Exploring the educational potential of robotics in schools: A systematic review. *Comput. Educ., 58*, 978–988.

13. Bishop, C. M., (2006). *Pattern Recognition and Machine Learning (Information Science and Statistics)*, (1st edn., 2006 corr. 2nd printing edn., Vol. 2, pp. 1–20). Springer.
14. Bosilj, P., Duckett, T., & Cielniak, G., (2018). Connected attribute morphology for unified vegetation segmentation and classification in precision agriculture. *Comput. Ind., 98*, 226–240.
15. Chan, T. F., & Vese, L. A., (2001). Active contours without edges. *IEEE Trans. Image Process, 10*, 266–277.
16. Chen, S. W., Shivakumar, S. S., Dcunha, S., Das, J., Okon, E., Qu, C., Taylor, C. J., & Kumar, V., (2017). Counting apples and oranges with deep learning: A data-driven approach. *IEEE Robot. Autom. Lett., 2*, 781–788.
17. Cheng, H., Damerow, L., Sun, Y., & Blanke, M., (2017). Early yield prediction using image analysis of apple fruit and tree canopy features with neural networks. *J. Imaging, 3*, 1–6.
18. Choi, D., Lee, W. S., Schueller, J. K., Ehsani, R., Roka, F., & Diamond, J., (2017). A performance comparison of RGB, NIR, and depth images in immature citrus detection using deep learning algorithms for yield prediction. In: *Proceedings of the 2017 Spokane* (Vol. 1, pp. 1–25). Washington, DC, USA.
19. Chuanlei, Z., Shanwen, Z., Jucheng, Y., Yancui, S., & Jia, C., (2017). Apple leaf disease identification using genetic algorithm and correlation-based feature selection method. *Int. J. Agric. Biol. Eng., 10*. doi: 10.3965/j.ijabe.20171002.2166.
20. Dias, P. A., Tabb, A., & Medeiros, H., (2018). Apple flower detection using deep convolutional networks. *Comput. Ind., 99*, 17–28.
21. Dorj, U. O., Lee, M., & Yun, S., (2017). An yield estimation in citrus orchards via fruit detection and counting using image processing. *Comput. Electron. Agric., 140*, 103–112.
22. Duan, L., Huang, C., Chen, G., Xiong, L., Liu, Q., & Yang, W., (2015). Determination of rice panicle numbers during heading by multi-angle imaging. *Crop J., 3*, 211–219.
23. Ebrahimi, M. A., Khoshtaghaza, M. H., Minaei, S., & Jamshidi, B., (2017). Vision-based pest detection based on SVM classification method. *Comput. Electron. Agric., 137*, 52–58.
24. García, J., Pope, C., & Altimiras, F., (2017). A distributed means segmentation algorithm applied to *Lobesia botrana* recognition.

Complexity, 1–14.

25. Garcia-Lamont, F., Cervantes, J., López, A., & Rodriguez, L., (2018). Segmentation of images by color features: A survey. *Neurocomputing, 292*, 1–27.
26. Hamuda, E., Mc Ginley, B., Glavin, M., & Jones, E., (2017). Automatic crop detection under field conditions using the HSV color space and morphological operations. *Comput. Electron. Agric., 133*, 97–107.
27. Han, S., Zhang, Q., Ni, B., & Reid, J., (2004). A guidance directrix approach to vision-based vehicle guidance systems. *Comput. Electron. Agric., 43*, 179–195.
28. He, B., Liu, G., Ji, Y., Si, Y., & Gao, R., (2011). Auto recognition of navigation path for harvest robot based on machine vision. In: *IFIP Advances in Information and Communication Technology* (Vol. 1, pp. 138–148). Springer: Berlin/Heidelberg, Germany.
29. Hiremath, S., Van, E. F. K., Braak, C. T., Stein, A., & Van, D. H. G., (2014). Image-based particle filtering for navigation in a semi-structured agricultural environment. *Biosyst. Eng., 121*, 85–95.
30. Hlaing, C. S., & Zaw, S. M. M., (2017). Plant diseases recognition for smart farming using model-based statistical features. In: *Proceedings of the 2017 IEEE 6th Global Conference on Consumer Electronics (GCCE)* (Vol. 1, pp. 1–4). Nagoya, Japan.
31. Hu, Q., Tian, J., & He, D., (2017). Wheat leaf lesion color image segmentation with improved multichannel selection based on the Chan-Vese model. *Comput. Electron. Agric., 135*, 260–268.
32. Iqbal, Z., Khan, M. A., Sharif, M., Shah, J. H., Ur Rehman, M. H., & Javed, K., (2018). An automated detection and classification of citrus plant diseases using image processing techniques: A review. *Comput. Electron. Agric., 153*, 12–32.
33. Islam, M., Anh, D., Wahid, K., & Bhowmik, P., (2017). Detection of potato diseases using image segmentation and multiclass support vector machine. In: *Proceedings of the 2017 IEEE 30th Canadian Conference on Electrical and Computer Engineering (CCECE)* (Vol. 1, pp. 1–4). Windsor, ON, Canada.
34. Ji, B., Zhu, W., Liu, B., Ma, C., & Li, X., (2009). Review of recent machine-vision technologies in agriculture. In: *Proceedings of the 2009 Second International Symposium on Knowledge Acquisition and Modeling* (; Vol. 1, pp. 330–334.). Wuhan, China.

35. Jiang, G. Q., Zhao, C. J., & Si, Y. S., (2010). A machine vision-based crop rows detection for agricultural robots. In: *Proceedings of the 2010 International Conference on Wavelet Analysis and Pattern Recognition, Qingdao* (Vol. 1, pp. 114–118). China.

36. Jiang, G., & Zhao, C., (2010). A vision system based crop rows for agricultural mobile robot. In: *Proceedings of the 2010 International Conference on Computer Application and System Modeling (ICCASM 2010)* (Vol. 1, pp. 142–145). Taiyuan, China.

37. Jiang, H., Xiao, Y., Zhang, Y., Wang, X., & Tai, H., (2013). Curve path detection of unstructured roads for the outdoor robot navigation. *Math. Comput. Model.58*, 536–544.

38. Kamilaris, A., & Prenafeta-Boldú, F. X., (2018). Deep learning in agriculture: A survey. *Comput. Electron. Agric., 147*, 70–90.

39. Kitchenham, B., (2004). *Procedures for Performing Systematic Reviews* (Vol. 1, pp. 1–30) Joint Technical Report, Computer Science Department, Keele University (TR/SE0401) and National ICT Australia Ltd. (0400011T.1), 2004; ISSN 13537776.

40. Knoll, F. J., Czymmek, V., Poczihoski, S., Holtorf, T., & Hussmann, S., (2018). Improving efficiency of organic farming by using a deep learning classification approach. *Comput. Electron. Agric., 153*, 347–356.

41. Kounalakis, T., Triantafyllidis, G. A., & Nalpantidis, L., (2018). Image-based recognition framework for robotic weed control systems. *Multimed. Tools Appl., 77*, 9567–9594.

42. Krizhevsky, A., Sutskever, I., & Hinton, G. E., (2017). ImageNet classification with deep convolutional neural networks. *Commun. ACM, 60*, 84–90.

43. Lecun, Y., Bottou, L., Bengio, Y., & Haffner, P., (1998). Gradient-based learning applied to document recognition. *Proc. IEEE, 86*, 2278–2324.

44. Li, B., Long, Y., & Song, H., (2018). Detection of green apples in natural scenes based on saliency theory and Gaussian curve fitting. *Int. J. Agric. Biol. Eng., 11*, 192–198.

45. Li, D., Zhao, H., Zhao, X., Gao, Q., & Xu, L., (2017). Cucumber detection based on texture and color in greenhouse. *Int. J. Pattern Recognit. Artif. Intell., 31*, 1754016.

46. Liu, H., Lee, S. H., & Chahl, J. S., (2017). An evaluation of the contribution of ultraviolet in fused multispectral images for invertebrate

detection on green leaves. *Precis. Agric., 18*, 667–683.

47. Liu, M. Y., Tuzel, O., Ramalingam, S., & Chellappa, R., (2011). Entropy rate super pixel segmentation. In: *Proceedings of the CVPR 2011* (Vol. 1, pp. 2097–2104). Providence, RI, USA.
48. Lottes, P., & Stachniss, C., (2017). Semi-supervised online visual crop and weed classification in precision farming exploiting plant arrangement. In: *Proceedings of the 2017 IEEE/RSJ International Conference on Intelligent Robots and Systems (IROS)* (Vol. 1, pp. 5155–5161). Vancouver, BC, Canada.
49. Lottes, P., Behley, J., Milioto, A., & Stachniss, C., (2018). Fully convolutional networks with sequential information for robust crop and weed detection in precision farming. *IEEE Robot. Autom. Lett., 3*, 2870–2877.
50. Lu, H., Cao, Z., Xiao, Y., Li, Y., & Zhu, Y., (2016). Region-based color modeling for joint crop and maize tassel segmentation. *Biosyst. Eng., 147*, 139–150.
51. Ma, J., Du, K., Zhang, L., Zheng, F., Chu, J., & Sun, Z., (2017). A segmentation method for greenhouse vegetable foliar disease spots images using color information and region growing. *Comput. Electron. Agric., 142*, 110–117.
52. Mallat, S. G., (1989). A theory for multi resolution signal decomposition: The wavelet representation. *IEEE Trans. Pattern Anal. Mach. Intell., 11*, 674–693.
53. Marchant, J. A., (1996). Tracking of row structure in three crops using image analysis. *Comput. Electron. Agric., 15*, 161–179.
54. McCarthy, C. L., Hancock, N. H., & Raine, S. R., (2010). Applied machine vision of plants: A review with implications for field deployment in automated farming operations. *Intell. Serv. Robot., 3*, 209–217.
55. Meng, Q., Qiu, R., He, J., Zhang, M., Ma, X., & Liu, G., (2015). Development of agricultural implement system based on machine vision and fuzzy control. *Comput. Electron. Agric., 112*, 128–138.
56. Meyer, G. E., Hindman, T. W., & Laksmi, K., (1999). Machine vision detection parameters for plant species identification. In: Meyer, G. E., & DeShazer, J. A., (eds.), *Proceedings of the SPIE* (Vol. 1, pp. 327–335). Washington, DC, USA.
57. Milioto, A., Lottes, P., & Stachniss, C., (2018). Real-Time semantic

segmentation of crop and weed for precision agriculture robots leveraging background knowledge in CNNs. In: *Proceedings of the 2018 IEEE International Conference on Robotics and Automation (ICRA)* (Vol. 1, pp. 2229–2235). Brisbane, QLD.

58. Momin, M. A., Rahman, M. T., Sultana, M. S., Igathinathane, C., Ziauddin, A. T. M., & Grift, T. E., (2017). Geometry based mass grading of mango fruits using image processing. *Inf. Process. Agric., 4*, 150–160.
59. Muscato, G., Prestifilippo, M., Abbate, N., & Rizzuto, I., (2005). A prototype of an orange picking robot: Past history, the new robot and experimental results. *Ind. Robot-An Int. J., 32*, 128–138.
60. Oppenheim, D., Edan, Y., & Shani, G., (2017). Detecting tomato flowers in greenhouses using computer vision. *Int. J. Comput. Electr. Autom. Control Inf. Eng., 11*, 104–109.
61. Ortiz, J. M., & Olivares, M., (2006). A vision-based navigation system for an agricultural field robot. In: *Proceedings of the 2006 IEEE 3rd Latin American Robotics, Symposium* (Vol. 1, pp. 106–114). Santiago, Chile.
62. Otsu, N., (1979). A *Threshold Selection Method from Gray-Level Histograms*. IEEE: Piscataway, NJ, USA.
63. Pajares, G., García-Santillán, I., Campos, Y., Montalvo, M., Guerrero, J., Emmi, L., Romeo, J., et al., (2016). Machine-vision systems selection for agricultural vehicles: A guide. *J. Imaging, 2*, 34.
64. Patil, J. K., & Kumar, R., (2017). Analysis of content based image retrieval for plant leaf diseases using color, shape, and texture features. *Eng. Agric. Environ. Food, 10*, 69–78.
65. Patrício, D. I., & Rieder, R., (2018). Computer vision and artificial intelligence in precision agriculture for grain crops: A systematic review. *Comput. Electron. Agric., 153*, 69–81.
66. Pereira, C. S., Morais, R., & Reis, M. J. C. S., (2017). Recent advances in image processing techniques for automated harvesting purposes: A review. In: *Proceedings of the 2017 Intelligent Systems Conference (IntelliSys)* (Vol. 1, pp. 566–575). London, UK.
67. Pérez, D. S., Bromberg, F., & Diaz, C. A., (2017). Image classification for detection of winter grapevine buds in natural conditions using scale-invariant features transform, bag of features and support vector machines. *Comput. Electron. Agric., 135*, 81–95.

68. Ponce, J. M., Aquino, A., Millán, B., & Andújar, J. M., (2018). Olive-fruit mass and size estimation using image analysis and feature modeling. *Sensors, 18*, 2930.
69. Prakash, R. M., Saraswathy, G. P., Ramalakshmi, G., Mangaleswari, K. H., & Kaviya, T., (2017). Detection of leaf diseases and classification using digital image processing. In: *Proceedings of the 2017 International Conference on Innovations in Information, Embedded and Communication Systems (ICIIECS)* (Vol. 1, pp. 1–4).
70. Prasetyo, E., Adityo, R. D., Suciati, N., & Fatichah, C., (2017). Mango leaf image segmentation on HSV and YCbCr color spaces using Otsu thresholding. In: *Proceedings of the 2017 3rd International Conference on Science and Technology-Computer (ICST)* (Vol. 3, pp. 99–103). Yogyakarta, Indonesia.
71. Radcliffe, J., Cox, J., & Bulanon, D. M., (2018). Machine vision for orchard navigation. *Comput. Ind., 98*, 165–171.
72. Reid, J. F., Zhang, Q., Noguchi, N., & Dickson, M., (2000). Agricultural automatic guidance research in North America. *Comput. Electron. Agric., 25*, 155–167.
73. Rosten, E., Porter, R., & Drummond, T., (2010). Faster and better: A machine learning approach to corner detection. *IEEE Trans. Pattern Anal. Mach. Intell., 32*, 105–119.
74. Rublee, E., Rabaud, V., Konolige, K., & Bradski, G., (2011). ORB: An efficient alternative to SIFT or SURF. In: *Proceedings of the 2011 International Conference on Computer Vision* (Vol. 1, pp. 2564–2571). Barcelona, Spain.
75. Sabanci, K., Kayabasi, A., & Toktas, A., (2017). Computer vision-based method for classification of wheat grains using artificial neural network. *J. Sci. Food Agric., 97*, 2588–2593.
76. Sabzi, S., Abbaspour-Gilandeh, Y., & Javadikia, H., (2017). Machine vision system for the automatic segmentation of plants under different lighting conditions. *Biosyst. Eng., 161*, 157–173.
77. Salembier, P., Oliveras, A., & Garrido, L., (1998). Antiextensive connected operators for image and sequence processing. *IEEE Trans. Image Process, 7*, 555–570.
78. Scarfe, A. J., Flemmer, R. C., Bakker, H. H., & Flemmer, C. L., (2009). Development of an autonomous kiwifruit picking robot. In: *Proceedings of the 2009 4th International Conference on Autonomous*

Robots and Agents (Vol. 1, pp. 380–384). Wellington, New Zealand.

79. Sethy, P. K., Routray, B., & Behera, S. K., (2019a). Detection and counting of marigold flower using image processing technique. In: *Lecture Notes in Networks and Systems* (Vol. 1, pp. 87–93). Springer: Singapore.
80. Sethy, P. K., Routray, B., & Behera, S. K., (2019b). Advances in computer, communication and control. In: Biswas, U., Banerjee, A., Pal, S., Biswas, A., Sarkar, D., & Haldar, S., (eds.), *Lecture Notes in Networks and Systems* (Vol. 41, pp. 1–50). Springer Singapore: Singapore.
81. Shalal, N., Low, T., McCarthy, C., & Hancock, N., (2013). A review of autonomous navigation systems in agricultural environments. *Soc. Eng. Agric. Conf. Innov. Agric. Technol. Sustain. Futur., 1*, 1–33.
82. Sharif, M., Khan, M. A., Iqbal, Z., Azam, M. F., Lali, M. I. U., & Javed, M. Y., (2018). Detection and classification of citrus diseases in agriculture based on optimized weighted segmentation and feature selection. *Comput. Electron. Agric., 150*, 220–234.
83. Sharifi, M., & Chen, X., (2015). A novel vision-based row guidance approach for navigation of agricultural mobile robots in orchards. In: *Proceedings of the 2015 6th International Conference on Automation, Robotics and Applications (ICARA)* (Vol. 1, pp. 251–255). Queenstown, New Zealand.
84. Søgaard, H. T., & Lund, I., (2007). Application accuracy of a machine vision-controlled robotic micro-dosing system. *Biosyst. Eng., 96*, 315–322.
85. Subramanian, V., Burks, T. F., & Arroyo, A. A., (2006). Development of machine vision and laser radar based autonomous vehicle guidance systems for citrus grove navigation. *Comput. Electron. Agric., 53*, 130–143.
86. Torii, T., (2000). Research in autonomous agriculture vehicles in Japan. *Comput. Electron. Agric., 25*, 133–153.
87. Torres-Sospedra, J., & Nebot, P., (2011). A new approach to visual-based sensory system for navigation into orange groves. *Sensors*, *11*, 4086–4103.
88. Vassallo-Barco, M., Vives-Garnique, L., Tuesta-Monteza, V., Mejía-Cabrera, H. I., & Toledo, R. Y., (2017). Automatic detection of nutritional deficiencies in coffee tree leaves through shape and texture

descriptors. *J. Digit. Inf. Manag., 15*, 7–18.

89. Viola, P., & Jones, M. J., (2004). Robust real-time face detection. *Int. J. Comput. Vis., 57*, 137–154.
90. Wan, P., Toudeshki, A., Tan, H., & Ehsani, R., (2018). A methodology for fresh tomato maturity detection using computer vision. *Comput. Electron. Agric., 146*, 43–50.
91. Wang, Q., Zhang, Q., Rovira-Más, F., & Tian, L., (2011). Stereovision-based lateral offset measurement for vehicle navigation in cultivated stubble fields. *Biosyst. Eng., 109*, 258–265.
92. Wang, Z., Walsh, K., & Verma, B., (2017). On-tree mango fruit size estimation using RGB-D images. *Sensors, 17*, 2738.
93. Wang, Z., Wang, K., Yang, F., Pan, S., & Han, Y., (2018). Image segmentation of overlapping leaves based on Chan-Vese model and sobel operator. *Inf. Process. Agric., 5*, 1–10.
94. Warner, M. G. R., & Harries, G. O., (1972). An ultrasonic guidance system for driverless tractors. *J. Agric. Eng. Res., 17*, 1–9.
95. Wilson, J., (2000). Guidance of agricultural vehicles—a historical perspective. *Comput. Electron. Agric., 25*, 3–9.
96. Xiong, X., Duan, L., Liu, L., Tu, H., Yang, P., Wu, D., Chen, G., Xiong, L., Yang, W., & Liu, Q., (2017). Panicle-SEG: A robust image segmentation method for rice panicles in the field based on deep learning and super pixel optimization. *Plant Methods, 13*, 100–104.
97. Xue, J., & Xu, L., (2010). Autonomous agricultural robot and its row guidance. In: *Proceedings of the 2010 International Conference on Measuring Technology and Mechatronics Automation* (Vol. 1, pp. 725–729). Changsha City, China.
98. Xue, J., Zhang, L., & Grift, T. E., (2012). Variable field-of-view machine vision-based row guidance of an agricultural robot. *Comput. Electron. Agric., 84*, 85–91.
99. Yaghoubi, S., Akbarzadeh, N. A., Bazargani, S. S., Bazargani, S. S., Bamizan, M., & Asl, M. I., (2013). Autonomous robots for agricultural tasks and farm assignment and future trends in agro robots. *Int. J. Mech. Mechatronics Eng., 13*, 1–6.
100. Yahata, S., Onishi, T., Yamaguchi, K., Ozawa, S., Kitazono, J., Ohkawa, T., Yoshida, T., Murakami, N., & Tsuji, H., (2017). A hybrid machine learning approach to automatic plant phenotyping for smart agriculture.

In: *Proceedings of the 2017 International Joint Conference on Neural Networks (IJCNN)* (Vol. 1, pp. 1787–1793). Anchorage, AK, USA.

101. Yang, W., Tang, W., Li, M., Zhang, D., & Zhang, Y., (2012). Corn tassel detection based on image processing. In: *Proceedings of the 2012 International Workshop on Image Processing and Optical Engineering* (Vol. 1, pp. 1–22). SPIE: Harbin.
102. Yukumoto, O., Matsuo, Y., & Noguchi, N., (2000). Robotization of agricultural vehicles (part 1)-Component technologies and navigation systems. *Japan Agric. Res. Q., 34*, 99–105.
103. Zhang, H., & Sha, Z., (2013). Product classification based on SVM and PHOG descriptor. *Int. J. Comput. Sci. Netw. Secur., 13*, 9–20.
104. Zhang, J., Kong, F., Zhai, Z., Wu, J., & Han, S., (2018). Robust image segmentation method for cotton leaf under natural conditions based on immune algorithm and PCNN algorithm. *Int. J. Pattern Recognit. Artif. Intell., 32*, 1854011.
105. Zhang, Q., Reid, J. J. F., & Noguchi, N., (1999). Agricultural vehicle navigation using multiple guidance sensors. *Proc. Int. Conf. F. Serv. Robot., 1*, 293–298.
106. Zhang, S., Wang, H., Huang, W., & You, Z., (2018). Plant diseased leaf segmentation and recognition by fusion of superpixel, K-means and PHOG. *Optik, 157*, 866–872.
107. Zhang, S., Wu, X., You, Z., & Zhang, L., (2017). Leaf image based cucumber disease recognition using sparse representation classification. *Comput. Electron. Agric., 134*, 135–141.
108. Zhou, R., Kaneko, S., Tanaka, F., Kayamori, M., & Shimizu, M., (2014). Disease detection of *Cercospora* leaf Spot in sugar beet by robust template matching. *Comput. Electron. Agric., 108*, 58–70.
109. Zuiderveld, K., (1994). Contrast limited adaptive histogram equalization. In: *Graphics Gems* (Vol. 1, pp. 474–485). Elsevier: Amsterdam, the Netherlands.

Chapter 5

Integration of Actuation Mechanisms in Agricultural Systems

CONTENTS

5.1. INTRODUCTION

In recent years, main advances have been taken place in the development of more powerful and new view systems for agriculture, such as global positioning systems and computer vision. An important stimulus has been received by the automation of agricultural tasks, due to these advances, particularly in the area where high precision is essential such as selective weed control for the proper use of the implementation and resources of more efficient treatments (Koyachi et al., 1995; Emmi et al., 2014). Perception systems have been integrated and incorporated by such autonomous agricultural systems for actuation systems that are accountable for acting the agricultural operations, decision-making systems for analyzing and interpreting such information, obtaining info from the environment (Figure 5.1).

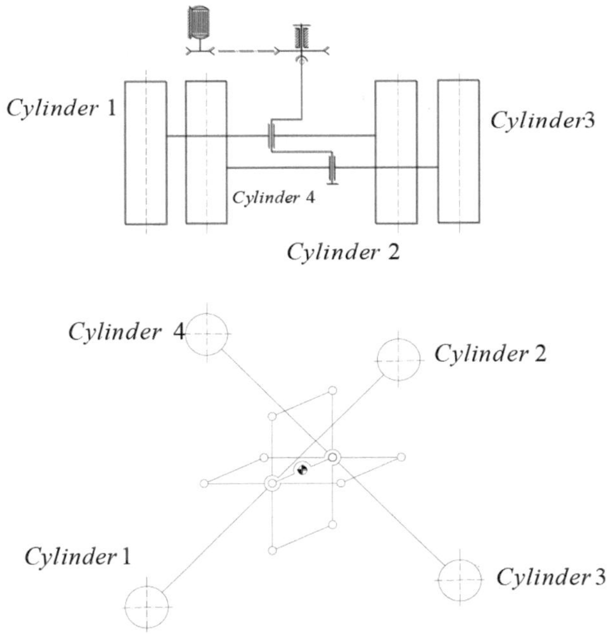

Figure 5.1: Basic components of actuation mechanism.

Source: https://www.researchgate.net/publication/292551454_STRUCTURAL_MODEL_FOR_AN_ACTUATION_MECHANISM_OF_PLANSIFTERS_IN_WHEAT_MILLS.

These systems contain different actuators, sensors, and computers that work synchronously in a precise style for the proposed aim (Lui et al.,

2009; Luka et al., 2015). The chief input of this chapter is the arrangement, selection, synchronization, and integration of these systems to make a whole independent vehicle for agricultural applications. This kind of vehicle has enticed rising interest, for researchers as well as for the farmers and manufacturers. The performance and success of the integrated system have been demonstrated in the experimental results in weed control tasks and guidance in a maize field, showing its efficiency and utility. In other agricultural tasks use, the whole system is adequately flexible, in the field of autonomous agricultural vehicles with little effort and is another important contribution (Goumopoulos et al., 2007, 2009).

In the 1920s, autonomous outdoor navigation of vehicles was proposed with actuation systems and integrated sensors, but in the 1980s it was realized first when the technology was developed enough to permit actual tests (Li et al., 2009). Presently in the field, there exists a rising interest with important progress (AuatCheein and Carelli, 2013). With the help of machine vision, NavLab was one of the more outstanding and first vehicles capable of navigating the dynamic and real environment, a heavy computing power, and range finder onboard the vehicle (Hebert and Kanade, 1986). A few years later, using different techniques and concepts some researchers tried to automate agricultural vehicles. Based on radio beacons, Erbach et al. (1991) proposed a static system to triangulate the vehicle's position for steering aims. To track a visual mark on the vehicle, an analogous system using cameras in the field was also used to control the position of the tractor (Noguchi and Terao, 1997). Researchers returned to the NavLab philosophy, although this system was successful by putting cameras onboard the vehicle. The static vision system changed toward mobile apparatus that was able to recognize its features and the environment for vehicle steering purposes. To track straight crop rows, by developing controllers for autonomous agricultural tractors, several researchers have been on this research by these developments (Billingsley and Schoenfisch, 1995; Gerrish et al., 1997).

At around the same time, based on (Global Positioning System (GPS), a different approach was proposed by O'Connor et al. (1996). These authors verified that how a carrier phase GPS equipped in an autonomous vehicle with four antennas can give both heading and position with accuracy high enough to achieve agricultural tasks in the field. Since then, for controlling and measuring a vehicle's heading and position, GPS has been accepted as the representative method, and some commercial systems have also been included in it (Rekow and Ohlemeyer, 2007). Nevertheless, new methods have recently been proposed using GPS for the autonomous guidance of tractors, and research is still ongoing (Kise and Zhang, 2008; Gomez-Gil et al., 2011).

To recognize environmental hindrances and particularities, machine vision has been revealed to be crucial, although for guiding agricultural vehicles GPS technology delivers good precision; hence, in the 2000s both methods started to be merged and for agricultural vehicles became the standard method (Stenz et al., 2002; Rovira-Más et al., 2003). Particularly for guidance, camera-based systems have been established as the main task for crop and weed discrimination, where guidance was significant (Jones et al., 2009; Zheng et al., 2009). Elements and sensors are mandatory for detection and guidance tasks to be suitably adjusted, decided, and calibrated on board the vehicle during operation for accuracy (Montalvo et al., 2012; Romeo et al., 2013).

To carry out farming tasks such as planting seeds, fertilizing, tilling, mowing, weed control, harvesting, and applying pesticides, and the agricultural system requires to be prepared with the proper implements, apart from the difficulties of making a vehicle autonomous. In the automation of agricultural systems, this is a step forward, which this book's authors are focused on. There are two main kinds of implements based on whether they are towed or carried by the vehicle. Positioning accuracy of the vehicle and the perception system are the factors that determine the positioning accuracy of the former type, and most present advances fall in this topic (Xue et al., 2012; Guerrero et al., 2013). The latter depends on the positioning accuracy of the vehicle, the positioning accuracy of the implement steering mechanism as well as the accuracy of the perception system, in positioning this type of system, some attempts have been made by using two GPS antennas, one located on the implement and other on the vehicle. For instance, an unmanned hoeing system has been developed by Nørremark et al. (2008) for intra-row weed management based on real-time kinematic (RTK) GPS methods. The system was composed of a side-shifting arrangement attached to weed implement and an autonomous vehicle. Both the implement and vehicle were equipped with RTK-GPS; hence, the two subsystems gave their position to permit implement to act on every plant and the vehicle to follow predefined GPS paths, whose positions were attained during seeding automatically. Based on the RTK-GPS perception system mechanical control effective actuation systems (cultivator and sprayer) have also been developed (Rovira-Más et al., 2010; Perez-Ruiz et al., 2013).

Productivity is one important concern in agriculture, where agricultural tasks have to be performed with minimal resources, maximum performance, and accuracy. This condition means that the incorporation of the above-mentioned systems (the implement and the vehicle) must be done under architecture with a reliable and effective design to satisfy all the requirements, particularly, the predictable real-time performance. Hence, the architecture is a vital subject, where all subsystems are to be organized. Significance of the active integration of computers, sensors, and actuators has been highlighted by Suprem et al. (2013). Great emphasis has been put on the establishment of the individual elements but not on their integration; note that integration is specifically important in agricultural vehicles where the suitable condition is to design open and flexible systems for more than one agricultural application, intending to make complete use of agricultural vehicles for an extensive range of agricultural applications as stated by Blackmore et al. (2005). Accordingly, hybrid agent-based architecture for agricultural purposes has been proposed by García-Pérez et al. (2008), where actuation and perception tasks are integrated and suitably coordinated (Qiu et al., 2001; Zolfagharian et al., 2016).

In autonomous robots, Slaughter et al. (2008) reviewed systems for agricultural tasks and recognized four main subsystems: precision actuation, guidance, mapping, and weed identification and detection. Weed identification and detection and guidance are based on imaging sensors and RTK-GPS. Weeding is a precise agricultural treatment based on electrocution and thermal devices, cutting tools, and micro-sprays where actuation systems are focused on precise control (Pedersen et al., 2008). Information obtained at a previous stage is applied to the application by a process called mapping; during sowing Bak and Jakobsen (2004) found a map that was used during the treatment of weeds.

Contrary to this, for intelligent autonomous vehicles, an open architecture was proposed by Rovira-Mas (2010a, b) based on three-layer (information, machine actuation, and safety) structure. All security aspects concerning the user's integrity and the vehicle are the responsibility of the safety layer. Processing of all data provided by the set of sensors onboard the tractor is the responsibility of the information layer (perception). Ultimately, the execution of decisions is carried out by the actuation layer (action and decision making) according to the intelligent processes. To achieve the problematic agricultural necessities, all of these layers are interrelated. It is essential to follow an architectural model, for the development of agricultural autonomous vehicles based on such systems with the mandatory scalability and flexibility to increase the range of the vehicle's applications while concurrently attaining satisfactory efficiency and robustness. Perception system in depth has been developed by Rovira-Mas (2010a) under a precise configuration highlighting the sensor placement for real-time purposes, along with investigating the subsequent four significant properties: robustness, flexibility, efficiency, and scalability (Zhang et al., 1999; Mayerle et al., 2006).

An architecture that integrates the above four subsystems have been proposed based on the above considerations (weed identification and detection, mapping, guidance, and precision actuation) while comprising the above-mentioned four properties (robustness, flexibility, efficiency, and scalability). This procedure is attained with the projected scheme showed in Figure 5.1. There are three main modules it has: decision making, acting, and sensing. Information collected from the environment is responsible for sensing through the set of sensors accessible (inertial systems, imaging, and GPS). For weed identification and detection, guidance, and mapping, the information must be suitable. According to tasks to be carried out, the sensors are adapted and new sensors when required could be added, such as for safety navigation range finders (flexibility). Each sensor can be substituted by similar sensors, depending on the agricultural application with different features (scalability). Sensors must determine the harsh environmental conditions (robustness). Data must be provided by all sensors to be synchronized for a real-time implementation (efficiency). Through specific algorithms and procedures, processing the information is the responsibility of the Decision-Making System for weed identification and detection, guidance, and mapping. To get all accessible information provided by the sensors, the hardware/software components are designed which can be deactivated /activated conveniently (scalability, flexibility). All processes and data are controlled by these components to ensure that they are received on time (robustness) with the essential coherence for real-time applications (efficiency). After this, choices are made to be transmitted (messages), when required, to the perception and/or actuation systems. Either on the agricultural implements or for navigation, control actions are applied for precise tasks, such as weeding. Diverse parts of the implements can be deactivated or activated (scalability and flexibility), and different implements should be possible. For site-specific applications, implements must act with the highest precision as possible (efficiency and robustness). All subsystems are connected with a suitable communication protocol. The main contribution of this chapter is the proposed architectural design with separated properties and functionalities. One of the main demands of farmers and manufacturers for different agricultural applications, open architecture permits the same robot to be used with little effort. In maize fields, this design has been tested successfully a high degree of performance (Chao, 2010; Chao and Chen, 2012) (Figure 5.2).

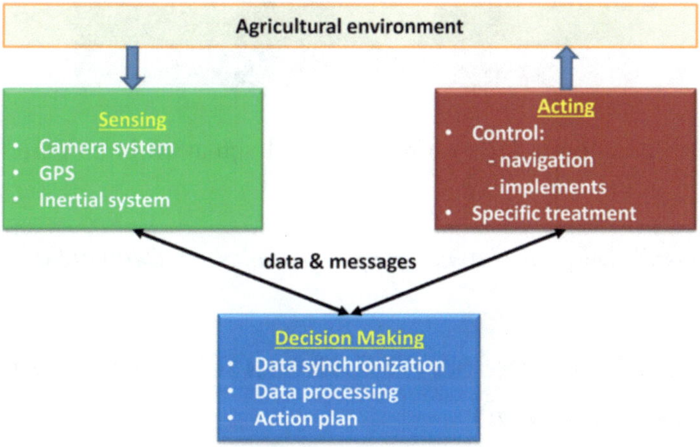

Figure 5.2: Architectural design: Perception (sensing), actuation, and decision making.

Source: https://www.researchgate.net/publication/260430835_Integrating_SensoryActuation_Systems_in_Agricultural_Vehicles.

Hence, this chapter aims to configure a whole S, perception, and decision making working on real wide-row crops as subsystems for an agricultural autonomous system (see Figure 5.3). In maize fields for weed control, a proposed architecture is designed (Peruzzi et al., 2012), but using the same vehicle with its architecture it should be suitable, for other agricultural purposes such as for seeding or cereal treatments (Hrazdera, 2005).

Figure 5.3: Perception, decision making, and actuation in an agricultural vehicle.

Source: https://www.ncbi.nlm.nih.gov/pmc/articles/PMC4003928/

5.2. PERCEPTION SYSTEM: LOCALIZATION AND WEED DETECTION

For crop/weed identification, the perception system is designed with two main goals: weed discrimination and row following for site-specific treatment. For both guidance weed discrimination, crop row detection is the basis and needs the identification and localization of the rows (straight line equations) in the image. Specification of the perception system, the properties, and the steps for data processing have been presented in this section (Frasconi et al., 2014; Ericson, 2017).

5.2.1. Description of the Perception System

There are three main sensors of the perception system: Inertial Measurement Unit (IMU), color camera, and RTK-GPS (comprising of two antennas: one for heading calculations and other for XYZ positioning). The main sensors on-board the vehicle has been displayed in Figure 5.4a based on a commercial tractor chassis. The IMU and camera are rooted in housing with a fan measured by a thermostat for cooling purposes, assuming that some agrarian jobs are accompanied under high working temperatures, above 50°C. To work in harsh environments, the housing is IP65 protected (exposure to drops of liquid from sprayers, dust, etc.). The perception system also includes a wireless communication device, additional to the appointed sensors, to enable the user to remotely monitoring and controlling the entire system. Figure 5.5 shows the assemblage of these parts (Romeo et al., 2012; Toh et al., 2014).

Figure 5.4: (a) Sensors on-board the tractor; (b) region of interest (ROI) for the vision system.

Source: https://www.ncbi.nlm.nih.gov/pmc/articles/PMC4003928/.

The camera-based sensor is from SVS-VISTEK with the SVS4050CFLGEA model and a GR Bayer color filter and is built with the CCD Kodak KAI 04050 M/C sensor; its resolution is 2,336 by 1,752 pixels with a 5.5 by 5.5 μm pixel size. Connected to the Main Controller, this camera is Gigabit Ethernet compliant. 3DM-GX3®-35 high-performance model miniature Attitude Heading Reference System (AHRS) with GPS is the IMU, of LORD MicroStrain® Sensing Systems (Williston, VT, USA). Via RS232, it is connected to the Main Controller and gives information about the roll angles and pitch.

For real-time performance, both the IMU and camera exhibit sufficient capabilities and are robust enough, needed for agricultural tasks. For precise heading calculation, the Trimble® BX982 GNSS receiver is the GPS receiver, supporting two antennas, mounted on the lateral ends of the vehicle (see Figure 5.4a). The aim is to apply precise treatments in front of the tractor, in the Region of Interest (ROI) which is a rectangular area 2 m long and 3 m wide (see Figure 5.4b). Four crop rows in the field are covered. Concerning a virtual vertical axis, this area starts at 3 m traversing the center of the image plane in the camera, i.e., where the scene is imaged (Bossu et al., 2009; Potena et al., 2016).

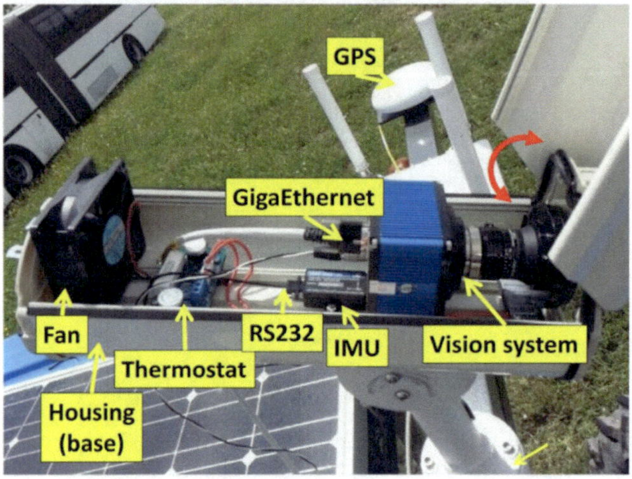

Figure 5.5: Perception system: sensors and equipment.

Source: https://www.researchgate.net/publication/260430835_Integrating_SensoryActuation_Systems_in_Agricultural_Vehicles.

Figure 5.6, displays the camera system geometry. *OXYZ* is the reference frame located in the ground with its axes oriented as displayed, aligned with the reference axis of the GPS; h is the height from *O* to the origin *o* of the reference frame *oxyz* attached to the camera; roll (θ), pitch (α), and yaw (β) define the three degrees of freedom of the image plane concerning the referential system; d is the distance from the beginning of the ROI measured perpendicular to the X-axis.

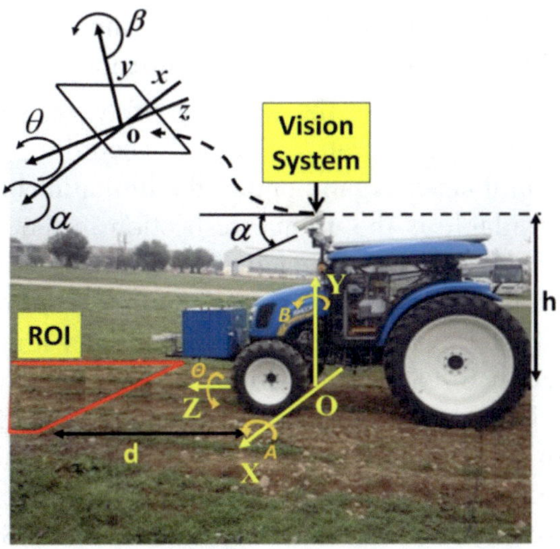

Figure 5.6: Camera system geometry.

Source: https://www.researchgate.net/publication/260430835_Integrating_SensoryActuation_Systems_in_Agricultural_Vehicles.

5.2.2. Characteristics of the Perception System

For the perception system, four properties have been recognized as elementary necessities united into the suggested architectural design:

- **Flexibility/Modularity:** The elements described above combine to make a perception system having a direct link of communication to the Decision-Making System consecutively on the Main Controller. Via RS232 and Giga-Ethernet communication ports, various sensors can be linked, which are standard interfaces. Without restrictions, any sensor can be connected/disconnected excluding the physical capacity of the Main Controller (see Section 4). Via software, the operation of decoupling, and linkage can be carried out, without upsetting the operability or the outstanding modules. According to the necessary agricultural tasks, a multi-sensor arrangement is allowed in this procedure. IMU is not needed in some crop line detection algorithms; this sensor is simply ignored when this happens, and then activated when essential. Significant disruptions to the designed systems are avoided by this method and at the same time, modularity, and flexibility of the proposed architecture are proved.

- **Scalability:** To raise the quantity of work, sensors, and their equivalent drivers can be added according to the required agricultural tasks. Again, one of the limitations of the number of ports available is the unique restriction in the Main Controller, which can be expanded easily. So, for plant discrimination, we could add a thermal camera or a multispectral or a stereoscopic vision-based system for safety purposes with a Giga-Ethernet connection for object detection. When needed various cameras with higher resolutions are accepted. The Decision-Making System implemented in the Main Controller permits for high-performance processing, regarding the processing of data delivered by the sensors (Slaughter et al., 2008; Midtiby et al., 2016).
- **Robustness:** The adverse outdoor agricultural environments are supported by a robust camera-based system with sufficient electronic and physical protection. It is intended to resist soil roughness, high variability in illumination, mechanical vibrations from the tractor's engine, and extreme temperatures. Moreover, to cut spectral infrared and ultraviolet radiation the camera is equipped with a UV/IR filter which significantly disturbs the quality of an image. To provide stable values, the IMU is calibrated and encapsulated (Gée et al., 2008; Burgos-Artizzu et al., 2011).
- **Real-Time/Performance:** The perception system is organized onboard the tractor, chiefly the camera, to cover the district of interest where the precise actuation is to be applied to minimize undesired mechanical effects and vibrations, it is placed close to the center of gravity. Through the exposure time, the image acquisition is controlled, again to attain high-quality images. Under specific modules, the data processing for crop/weed guidance and detection is designed, embedded in the Main Controller, and programmed as dynamic-link libraries (DLL) in C++ (Gongal et al., 2015; Dimililer and Kiani, 2017). For the planned specific treatment, these modules are adjusted to work in real-time.

5.2.3. Process: Integration of Information

A general mission or a pre-defined plan is followed by the program fed in the vehicle (see Figure 5.7). Based on GPS coordinates, a set of waypoints are established that are contained in the mission, which describes the end positions (target points) and the beginning (starting points) for crossing the field. Some uncertainty is present in the information from the mission, for this particular case, so that the endpoints and start point may not match the matching crop line's center. The use of the crop row detection system is required to correct this uncertainty existing in the general mission (Korthals et al., 2018; Gai et al., 2019).

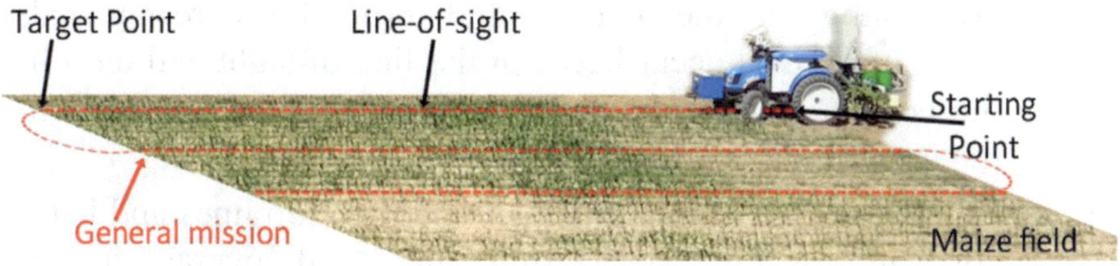

Figure 5.7: Example of path planning in the maize field.

Source: https://www.ncbi.nlm.nih.gov/pmc/articles/PMC4003928/.

The perception system carries info on the weed invasion in the field, in addition to the crop row detection, permitting the computation of the percentage of weed existence inside the ROI. The ROI is divided into rectangular sections, to do this, 0.25 m long (which agrees to the minimum resolution for the implementation of the treatment) and 0.375 m wide (agreeing to split in two the inter-row space, to classify the infestation on each side of the crop row). Based on the characteristics of the actuation system, the size of these sections is attuned that does the treatment. The process is as follows (Wang et al., 2019):

1. For this type of treatment, the operation speed set is defined as 0.83 m/s (3 km/h);
2. The traversal order of the waypoints to be visited is determined by the pre-defined plan, including target and starting points;
3. The vehicle follows the line-of-sight, between two waypoints inside the field;
4. The camera captures images up to ten images per second at frame rates;

- The system captures an image whenever the vehicle moves within 2 m on the field and reads the GPS coordinates at a rate of 10 Hz which is the length of the ROI;
- To recognize four crop lines, the camera vision system processes each image. Information about extrinsic camera parameters is provided by the IMU, roll (θ), and pitch (α) so that four expected crop lines are identified together with remaining intrinsic and extrinsic parameters. To determine the real crop lines, the expected crop lines serve as guidelines;
- Concerning the center of the image, based on the relative positioning of the two central crop lines recognized, if nonconformities occur between the line-of-sight and crop lines, the heading and the lateral deviation are corrected to align the tractor with the real crop lines in the field; and
- Based on the green densities around the crop lines and between adjacent crop lines, to determine the weed coverage inside the ROI, the detected crop lines are used.

The image analysis process consumes a precise amount of time, because the raw image received from the camera is conveyed by Ethernet (non-deterministic procedure), and the actuation point (implement elements that perform the treatment) is situated some meters behind the ROI, to maintain a high-accuracy treatment and a real-time performance an element for georeferencing and synchronization is needed. Each image captured is associated with a GPS position, to accomplish this synchronization and a map for weed coverage is shaped that is referenced to GPS coordinates.

5.3. ACTUATION/SENSING IN CROPS

Actuation in crops is usually performed by devices or implements towed/ pulled by tractors that provide motion and energy, which is the primary aim in agriculture. To obtain the status of the crop, implements are used which are tools composed of simple sensors and simple actuators to achieve simple actuation such as moving prismatic cylinders, opening/closing valves, etc. Most of these actuators/ sensors are measured by using PLCs or similar devices. If sophisticated sensor systems such as range finders or vision systems are essential, as indicated in the previous section they are provided as external devices (Peteu et al., 2010; Dicker et al., 2014).

5.3.1. Description of the Actuation System

Although the proposed system can be applied to a large number of implements, we will specify the system for a particular mechanical-thermal machine, which is devoted to weed control (see Figure 5.8a) in flame-resistant crops such as maize, onion, and garlic. This implement is pulled by the autonomous tractor, and the Main Controller is in charge of decision making and synchronizing the activation of the treatment as well as managing the lateral position of the implement concerning the vehicle's position (Guo et al., 2015; Li and Wang, 2016).

Figure 5.8: (a) Mechanical-thermal machine; (b) Implement steering actuator (implement front view); (c) Main frame (implement rear view).

Source: https://www.ncbi.nlm.nih.gov/pmc/articles/PMC4003928/.

To address four succeeding crop rows four pairs of burners (see Figure 5.7a) of about 0.25 m long, attached to the mainframe are included in specific implement. The discriminating treatment in the intra-row space is the objective of the burners. The mechanical actuation, i.e., particular hoes attains the treatment in the inter-row space (see Figure 5.8a). The weed recognition system presented in previous sections recognizes the weed treatment on which every burner's flame concentration is dependent (Frasconi et al., 2012). The Actuation Controller which obtains the action mails from the Main Controller accomplishes the control of the ignition of the burners (Pokroy et al., 2009; Jiang et al., 2014).

It is prearranged that there are two different elements of implement to achieve weed control, and a suitable degree of precision in positioning the burners and mechanical elements possible risk of crop damage is required. The guidance system achieves this step, which is responsible for executing small adjustments concerning the vehicle in the lateral position of the implement. The linear actuator (central double rod hydraulic cylinder, Figure 5.8b) that permitting the operative machine to move across relating to the automobile and alters the angle of the steering wheels, is included in lateral positioning system (Figure 5.8c). This cylinder that is motorized by the hydraulic system of the vehicle, having a displacement of ±0.031 and it releases a force of 2,000 N. Through the controller area network (CAN) bus communication system onboard the autonomous vehicle Main Controller performs the lateral control of the implement (Wang et al., 2018) (Figure 5.9).

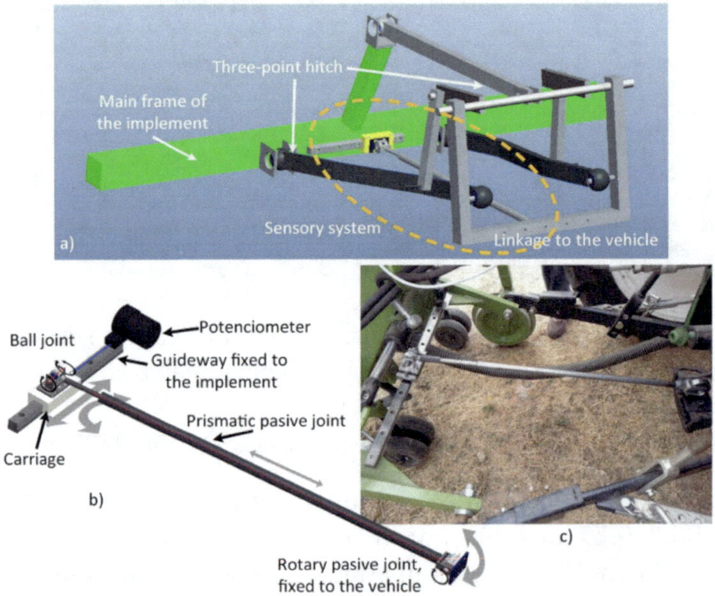

Figure 5.9: Lateral-position sensor setup. (a) Assembly diagram of the sensory system between the implement and the vehicle; (b) Structure of the sensory system; (c) Photo of the actual sensory system.

Source: https://www.researchgate.net/publication/260430835_Integrating_SensoryActuation_Systems_in_Agricultural_Vehicles.

5.3.1.1. Lateral-Position Sensor

A passive mechanism and a positioning sensor are combined to make the standing device for calculating the horizontal displacement of the implement regarding the vehicle. The passive mechanism depends on a telescopic arm that joins the implement mainframe to the vehicle's rear (see Figure 5.9b). With the rotation axis vertical to the arm, an inert rotating joint messes the end of the arm with the vehicle. Via a ball joint the carriage and the other end fixes. The linear guide provides a place for carriage to slip over. The dislocation of the implement can be calculated by the sensor as the cable-pull potentiometer (JX-PA-20-N14-13S-125) is attached to the carriage. Therefore, in a direction vertical to the vehicle's longitudinal axis, the dislocation of the implement about the vehicle is calculated by the positioning device. The basic structure of the sensory system is shown in Figure 5.9b, and Figure 5.9a, c presented its placement between the implement and the vehicle. For the alteration of the angle of the steering wheels for more accurate control of the implement guidance, in a hydraulic cylinder, an encoder (JX-EP-20-N14-110-25C) was positioned along with the lateral-position sensor for measuring the displacement of the implement about the vehicle.

5.3.1.2. Measuring the Implement Lateral Position Concerning the Vehicle

The real cross displacement of the implement relays with potentiometer dimensions based on the three-point holdup geometric model and the sensory system setup. For joining, the implement there is a diagram model of the three-point hitch presented in Figure 5.9, where the difference in the potentiometer measurement is presented as Δd, and ΔX is the real implement dislocation regarding the center of the vehicle. Regarding the central position axis h the locations of the points Pl and Pr are represented by Eq. (1) and Eq. (2) based on the geometric model of the three-point hitch (Figure 5.10). The position of the point Pr as a function of θ is represented by Eq. (3) (Figure 5.10), in which the excursion angle of the three-point hitch left arm is represented. As a function of Pr and Pl, the implement center axis I is represented by Eq. (4). It can be calculated, given the geometric model of the link between the implement and the three-point hitch and knowing the dimensions of the range and the arms of θ (Figure 5.10), the real displacement ΔX, the case of implement maximum excursion described in Eq. (5) is 0.17% greater than described in Eq. (7), the measured displacement Δd, which is negligible.

$$(X_{Pl} - (Xh - a/2))^2 + (Y_{Pl} - Yh)^2 = L^2 \quad (1)$$

$$(X_{Pr} - (Xh + a/2))^2 + (Y_{Pr} - Yh)^2 = L^2 \quad (2)$$

$$(X_{Pl}, Y_{Pl}) = \left((Xh - a/2) - L \cdot \cos\theta, Yh - L \cdot \sin\theta\right) \quad (3)$$

$$(Xi, Yi) = (Pl + Pr)/2 \quad (4)$$

$$\Delta X = Xh - Xi \quad (5)$$

$$\Delta Y = (Yh - \left(\sqrt{1 - D^2}\right) \cdot L - Yi \quad (6)$$

$$\Delta d = \sqrt{\Delta X^2 + \Delta Y^2} \quad (7)$$

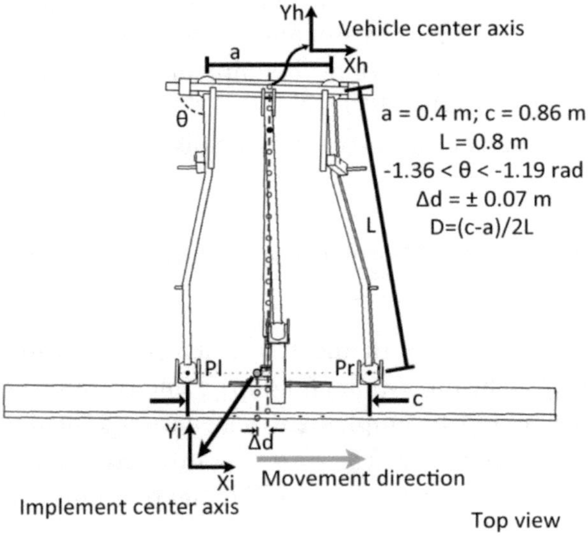

Figure 5.10: Schematic model of the three-point hitch for connecting the implement.

Source: https://www.ncbi.nlm.nih.gov/pmc/articles/PMC4003928/.

5.3.2. Characteristics of the Actuation System

Following are the properties defined for the perception system in Section 2, the actuation system exhibits the following characteristics:

- **Scalability:** The number of that can be connected depends on provided by the Main Controller provides the number of I/O channels that have a dependence on the connection of a large number of electrical, hydraulic, or pneumatic actuators (relays, prismatic cylinders, etc.). The I/O channels that can be even cataract-connected, which varieties the number of I/O ports nearly countless is used in large quantities by the designer whose permission is given by a particular controller used in this work (see Section 4). However, Scalability is just a slight dilemma.

- **Flexibility/Modularity:** Major border and simple actuators are included in the actuation system that pertains a given process to the crop. In this way, for changed crops, the particular actuators (burners, nozzles, etc.) can be easily changed and the major frame and the related placing systems can be utilized for a large number of crops. These machines can use actuators of several various types, as electric and hydraulic power is typically offered by agricultural vehicles. In this way for the treatment, the actuation system has a slight dependency on the nature of the crop.
- **Robustness:** As they consist of a made of steel and several commercial actuators make the mainframe that is used by actuation systems that are sufficiently rough for use in industrial and natural environments, high robustness is showed by the system.
- **Real-Time/Performance:** The PID controller that does not necessitate high computing power is needed in lateral positioning of the mainframe, and a CAN bus with few delays normally trigger the signals. Though risking the control performance, hydraulic valves can have a gradual response. Therefore, there is serious and must be cautiously planned real-time performance in actuation systems.

5.3.3. Process: Integration of Information

Two major tasks including (a) regarding the vehicle the lateral positioning of the mainframe is controlled and (b) controlling the beginning/ending of the burners are the parts of the actuation system. The action must be functional based on a weed coverage matrix is the first data which the vision system (camera-based sensor and processing) is conveyed directly The tool must be a few seconds later is designated by vehicle's speed, the position regarding the crop lines, and the changing of the heading, is the second information that is given by location system (GPS).

5.4. PERCEPTION-ACTUATION INTEGRATION AND DECISION MAKING

The Main Controller runs the Decision-Making System which connects and integrates the perception system. Furthermore, Decision-Making System also links the part of the actuation system responsible for implement lateral control. To assistant the same reference system for every chunk of information, the Decision-Making System is primarily in charge of coordinating the information coming from the different sensors. Therefore, the association is generated between where and when the actuation is needed and what is perceived (Niemueller et al., 2012; Ye et al., 2020).

Other significant responsibilities which are as follows: (a) the supervision of the development of the general mission; (b) the interpretation of the information; (c) the creation of an action plan to be executed by the actuation system; and (d) the evaluation of its reliability are in charge by Decision-Making System. The conceptual structure of interaction between the perception system, the Decision-Making System, and the actuation system is presented in the form of structure in Figure 5.1 (Shrobe et al., 2006; Scaramuzza et al., 2014).

5.4.1. Description of the Decision-Making Systems

The information that the sensors give about creating the actuation plan, the environment, and checking the implementation of the many tasks that make up the general mission has a dependence on the Decision-Making System. With the aim of actuations over agricultural fields for the managers and actuators, the judgment is made by understood, coordinated, handled, and arranged the information. To mandate certain actions on the respective devices messages are sent to the sensors and actuators by this system. Consisting of an LX150 FPGA with a Real-Time Operating System, CompactRIO-9082, with a 1.33 GHz dual-core Intel Core i7 processor includes in Main Controller, in which all these processes run. As the growing environment, LabVIEW Real-Time from National Instruments, published 2011 is used.

The superiority of the developed information may be influenced by some disruptions which happen by the environmental interaction, though in the perception system, the commercial devices are very reliable. Disturbances in the copy due to expressions or lighting changes, a reduction in the GPS accuracy as well as the loss of a message are some of the disturbances. In real-time, these disturbances can detect, and an approximation of the consistency of such information can be made by the grade of the impact of the disturbance (Chandra and Rocha, 2016; Niemueller et al., 2016).

The accuracy with which at that instant in time the action is performed is affected by the loss, failure, or alteration of information, given that this movement is satisfying the treatment and the vehicle is in motion. For example, instead of coming back to that particular area, it is better to operate the treatment with the worst-case situation in the case of low reliability in the generation of a weed cover map in a small section of the mission,

5.4.2. Characteristics of the Decision-Making System

On the basis above deliberations, Chandra and Rocha (2016) proposes a design that meets the above requirements under Lab view, where for architectural design in agricultural robots can be brief four properties as follows:

- **Flexibility/Modularity:** When it is obligatory, the data obtains/ directs from/to the perception and actuation systems by the Decision-Making System. The respective devices are linked directly to them, and in LabVIEW, these data are plotted as variables. SubVIs are the functions in the form of which all processing and control modules are written and they are combine related. The image processing module is a DLL established in C++ and is also in print as a subVI. It is meant that conferring to the particular requirements of each agricultural application, there can be an easy replacement, addition, and removal of all modules.
- **Scalability:** By adding new functionalities implanted as new modules and associate with internal links, the system can grow up simply so architectural design reaches a high level of scalability.
- **Robustness:** With access protection IP20 and effective humidity up to 90%, a high range of working temperatures (0°C to +55°C) is favored by RIO. Operational sinusoidal vibrations up to 500 Hz at 5 G is reinforced by this computer. The real responsibilities with exceptional strength can be achieved by the system and this system has been analyzed in real, tough agricultural conditions. Different robotics systems having independent agricultural robots have analyzed Lab VIEW effectively. The outer connection of communication is repressed to gain part of this feature in the control of the different devices between the different components.

The methodical data coming from the different system is settled by local protocols.

- **Real-Time/Performance:** Underneath a real-time operating system, for competing Lab VIEW is expressly created. Through Wi-Fi, where the Wi-Fi network recognizes restrictions, the remote communications are permitted by it and not via Lab VIEW. For broadcasting, this program is appropriate with a base station.

5.4.3. Process: Integration of Information

There are three main tasks including (a) weed coverage map interpretation (b) perception and actuation synchronization (c) trajectory planning are followed by Decision-Making System.

The overall objective is performed by Decision-Making System and to encounter the most optimal execution (shortest distance) the strategy centered on that task is accomplished. From starting and the last point, straight or curved lines are tracked by vehicle is included in this panning. The RTK-GPS that is modified by the camera-based system when required is the key sensory system which is used to close the control loop for way tracing. The way tracing has dependence on Decision-Making System through which planned sub-path is sent to the actuation system.

Sideways every sub-path, to develop reflections in synchronization with the steering velocity the control is given by Decision-Making System to perception system Ahead of the tractor every two meters is generated by every order all along each sub-path. The length of the ROI fixes this magnitude, and GPS offers the exact positions. To prevent gaps and uncovered areas, there should be an assurance of Full area coverage (Gautam et al., 2018).

REFERENCES

1. AuatCheein, F. A., & Carelli, R., (2013). Agricultural robotics: Unmanned robotic service units in agricultural tasks. *IEEE Ind. Electron. Mag., 7*, 48–58.
2. Bak, T., & Jakobsen, H., (2004). Agricultural robotic platform with four wheel steering for weed detection. *Biosyst. Eng., 87*, 125–136.
3. Billingsley, J., & Schoenfisch, M., (1995). Vision-guidance of agricultural vehicles. *Auton. Robots, 2*, 65–76.
4. Blackmore, B. S., Stout, W., Wang, M., & Runov, B., (2005). Robotic Agriculture—The future of agricultural mechanisation? In: *Proceedings of the 5th European Conference on Precision Agriculture* (Vol. 5, pp. 621–628). Uppsala, Sweden.
5. Bossu, J., Gée, C., Jones, G., & Truchetet, F., (2009). Wavelet transform to discriminate between crop and weed in perspective agronomic images. *Computers and Electronics in Agriculture, 65*(1), 133–143.
6. Burgos-Artizzu, X. P., Ribeiro, A., Guijarro, M., & Pajares, G., (2011). Real-time image processing for crop/weed discrimination in maize fields. *Computers and Electronics in Agriculture, 75*(2), 337–346.
7. Chandra, R., & Rocha, R. P., (2016). Knowledge-based framework for human-robots collaborative context awareness in usar missions. In: *2016 International Conference on Autonomous Robot Systems and Competitions (ICARSC)* (Vol. 1, pp. 335–340). IEEE.
8. Chao, H., & Chen, Y., (2012). *Remote Sensing and Actuation Using Unmanned Vehicles* (Vol. 3, pp. 1–32). John Wiley & Sons.
9. Chao, H., (2010). Cooperative remote sensing and actuation using networked unmanned vehicles. *All Graduate Theses and Dissertations, 1*, 590–597.
10. Dicker, M. P. M., Rossiter, J. M., Bond, I. P., & Weaver, P. M., (2014). Biomimetic photo-actuation: Sensing, control and actuation in sun-tracking plants. *Bioinspiration and Biomimetics, 9*(3), 036015.
11. Dimililer, K., & Kiani, E., (2017). Application of back propagation neural networks on maize plant detection. *Procedia Computer Science, 120*, 376–381.
12. Emmi, L., Gonzalez-de-Soto, M., Pajares, G., & Gonzalez-de-Santos, P., (2014). Integrating sensory/actuation systems in agricultural vehicles. *Sensors, 14*(3), 4014–4049.

13. Erbach, D. C., Choi, C. H., & Noh, K., (1991). Automated Guidance for Agricultural Tractors. In: *Automated Agriculture for the 21st Century* (Vol. 1, pp. 1–33). ASAE: Washington, DC, USA.
14. Ericson, S., (2017). *Vision-Based Perception for Localization of Autonomous Agricultural Robots* (Vol. 1, pp. 1–33).
15. Frasconi, C., Fontanelli, M., Raffaelli, M., Martelloni, L., & Peruzzi, A., (2012). Sensor based LPG management system for application of precision flaming. In: *Proceedings of the 1st International Conference on Robotics and Associated High-technologies and Equipment for Agriculture* (Vol. 1, pp. 67–72). RHEA-2012, Pisa, Italy.
16. Frasconi, C., Martelloni, L., Fontanelli, M., Raffaelli, M., Emmi, L., Pirchio, M., & Peruzzi, A., (2014). Design and full realization of physical weed control (PWC) automated machine within the RHEA project. In: *Proceedings of the International Conference of Agricultural Engineering* (Vol. 1, pp. 6–10). Zurich, Switzerland.
17. Gai, J., Tang, L., & Steward, B. L., (2019). Automated crop plant detection based on the fusion of color and depth images for robotic weed control. *Journal of Field Robotics, 1*, 1–32.
18. García-Pérez, L., Garcaí-Alegre, M. C., Ribeiro, A., & Guinea, D., (2008). An agent of behaviour architecture for unmanned control of a farming vehicle. *Comput. Electron. Agric.60*, 39–48.
19. Gautam, D., Watson, C., Lucieer, A., & Malenovský, Z., (2018). Error budget for geolocation of spectroradiometer point observations from an unmanned aircraft system. *Sensors*, *18*(10), 3465.
20. Gée, C., Bossu, J., Jones, G., & Truchetet, F., (2008). Crop/weed discrimination in perspective agronomic images. *Comput. Electron. Agric.60*, 49–59.
21. Gerrish, J. B., Fehr, B. W., Van, E. G. R., & Welch, D. P., (1997). Self-steering tractor guided by computer vision. *Appl. Eng. Agric.13*, 559–563.
22. Gomez-Gil, J., Alonso-Garcia, S., Gómez-Gil, F. J., & Stombaugh, T., (2011). A simple method to improve autonomous GPS positioning for tractors. *Sensors*, *11*, 5630–5644.
23. Gongal, A., Amatya, S., Karkee, M., Zhang, Q., & Lewis, K., (2015). Sensors and systems for fruit detection and localization: A review. *Computers and Electronics in Agriculture*, *116*, 8–19.

24. Gonzalez-de-Santos, P., Ribeiro, A., & Fernandez-Quitanilla, C., (2012). The RHEA Project: Using a robot fleet for a highly effective crop protection. In: *Proceedings of the International Conference of Agricultural Engineering* (Vol. 12, pp. 1–32). Valencia, Spain.
25. Goumopoulos, C., Kameas, A. D., & Cassells, A., (2009). An ontology-driven system architecture for precision agriculture applications. *International Journal of Metadata, Semantics and Ontologies*, *4*(1, 2), 72–84.
26. Goumopoulos, C., Kameas, A., & O'Flynn, B., (2007). Proactive agriculture: An integrated framework for developing distributed hybrid systems. In: *International Conference on Ubiquitous Intelligence and Computing* (Vol. 1, pp. 214–224). Springer, Berlin, Heidelberg.
27. Guerrero, J. M., Guijarrso, M., Montalvo, M., Romeo, J., Emmi, L., Ribeiro, A., & Pajares, G., (2013). Automatic expert system based on images for accuracy crop row detection in maize fields. *Expert Syst. Appl.*, *40*, 656–664.
28. Guo, Q., Dai, E., Han, X., Xie, S., Chao, E., & Chen, Z., (2015). Fast nastic motion of plants and bioinspired structures. *Journal of the Royal Society Interface*, *12*(110), 20150598.
29. Hebert, M., & Kanade, T., (1986). Outdoor scene analysis using range data. In: *Proceedings of the IEEE International Conference on Robotics and Automation* (Vol. 1, pp. 1426–1432). San Francisco, CA, USA.
30. Hrazdera, O., (2005). U.S. Patent No. 6,942,595, pp. 1–20. Washington, DC: U.S. Patent and Trademark Office.
31. Jiang, W., Wang, L., Liu, H., Ma, H., Tian, H., Chen, B., & Ding, Y., (2014). Bio-inspired directional high-aspect-ratio nanopillars: Fabrication and actuation. *RSC Advances*, *4*(79), 42002–42008.
32. Jones, G., Gée, C., & Truchetet, F., (2009). Assessment of an inter-row weed infestation rate on simulated agronomic images. *Comput. Electron. Agric.*, *67*, 43–50.
33. Kise, M., & Zhang, Q., (2008). Development of a stereovision sensing system for 3D crop row structure mapping and tractor guidance. *Biosyst. Eng.*, *101*, 191–198.
34. Korthals, T., Kragh, M., Christiansen, P., Karstoft, H., Jørgensen, R. N., & Rückert, U., (2018). Multi-modal detection and mapping of static and dynamic obstacles in agriculture for process evaluation. *Frontiers in Robotics and AI*, *5*, 28.

35. Koyachi, N., Arai, T., Adachi, H., Asami, K. I., & Itoh, Y., (1995). Hexapod with integrated limb mechanism of leg and arm. In: *Proceedings of 1995 IEEE International Conference on Robotics and Automation* (Vol. 2, pp. 1952–1957). IEEE.
36. Li, M., Imou, K., Wakabayashi, K., & Yokoyama, S., (2009). Review of research on agricultural vehicle autonomous guidance. *Int. J. Agric. Biol. Eng., 2*, 1–16.
37. Li, S., & Wang, K. W., (2016). Plant-inspired adaptive structures and materials for morphing and actuation: A review. *Bioinspiration and Biomimetics, 12*(1), 011001.
38. Lui, C., Cady, N. C., & Batt, C. A., (2009). Nucleic acid-based detection of bacterial pathogens using integrated micro fluidic platform systems. *Sensors, 9*(5), 3713–3744.
39. Luka, G., Ahmadi, A., Najjaran, H., Alocilja, E., DeRosa, M., Wolthers, K., & Hoorfar, M., (2015). Microfluidics integrated biosensors: A leading technology towards lab-on-a-chip and sensing applications. *Sensors, 15*(12), 30011–30031.
40. Mayerle, D. J., Engen, R. N., & Evans, A. W., (2006). U.S. Patent No. 7,140,310, pp. 1–20. Washington, DC: U.S. Patent and Trademark Office.
41. Midtiby, H. S., Åstrand, B., Jørgensen, O., & Jørgensen, R. N., (2016). Upper limit for context-based crop classification in robotic weeding applications. *Biosystems Engineering, 146*, 183–192.
42. Montalvo, M., Pajares, G., Guerrero, J. M., Romeo, J., Guijarro, M., Ribeiro, A., Ruz, J. J., & De La Cruz, J. M., (2012). Automatic detection of crop rows in maize fields with high weeds pressure. *Expert Syst. Appl., 39*, 11889–11897.
43. Niemueller, T., Lakemeyer, G., & Srinivasa, S. S., (2012). A generic robot database and its application in fault analysis and performance evaluation. In: *2012 IEEE/RSJ International Conference on Intelligent Robots and Systems* (Vol. 1, pp. 364–369). IEEE.
44. Niemueller, T., Neumann, T., Henke, C., Schönitz, S., Reuter, S., Ferrein, A., & Lakemeyer, G., (2016). International harting open source award 2016: Fawkes for the robocup logistics league. In: *Robot World Cup* (pp. 634–642). Springer, Cham.
45. Noguchi, N. K., & Terao, H., (1997). Development of an agricultural mobile robot using a geomagnetic direction sensor and image sensors.

J. Agric. Eng. Res., 67, 1–15.

46. Nørremark, M., Griepentrog, H. W., Nielsen, J., & Søgaard, H. T., (2008). The development and assessment of the accuracy of an autonomous GPS-based system for intra-row mechanical weed control in row crops. *Biosyst. Eng., 101*, 396–410.
47. O'Connor, M., Bell, T., Elkaim, G., & Parkinson, B., (1996). Automatic Steering of Farm Vehicles Using GPS. In: *Proceedings of the 3rd International Conference on Precision Agriculture* (Vol. 3, pp. 1–39). Minnesota, MI, USA.
48. Pedersen, S. M., Fountas, S., & Blackmore, S., (2008). Agricultural robots-applications and economic perspectives. In: Takahashi, Y., (ed.), *Service Robot Applications* (Vol. 1, pp. 369–382). In Tech: Rijeka, Croatia.
49. Perez-Ruiz, M., Carballido, J., Agüera, J., & Rodríguez-Lizana, A., (2013). Development and evaluation of a combined cultivator and band sprayer with a row-centering RTK-GPS guidance system. *Sensors, 13*, 3313–3330.
50. Peruzzi, A., Frasconi, C., Martelloni, L., Fontanelli, M., & Raffaelli, M., (2012). Application of Precision Flaming to Maize and Garlic in the RHEA Project. In: *Proceedings of the 1st International Conference on Robotics and Associated High-technologies and Equipment for Agriculture* (Vol. 1, pp. 55–60). RHEA-2012, Pisa, Italy.
51. Peteu, S. F., Oancea, F., Sicuia, O. A., Constantinescu, F., & Dinu, S., (2010). Responsive polymers for crop protection. *Polymers, 2*(3), 229–251.
52. Pokroy, B., Epstein, A. K., Persson-Gulda, M. C., & Aizenberg, J., (2009). Fabrication of bioinspired actuated nanostructures with arbitrary geometry and stiffness. *Advanced Materials, 21*(4), 463–469.
53. Potena, C., Nardi, D., & Pretto, A., (2016). Fast and accurate crop and weed identification with summarized train sets for precision agriculture. In: *International Conference on Intelligent Autonomous Systems* (Vol. 1, pp. 105–121). Springer, Cham.
54. Qiu, H., Zhang, Q., & Reid, J. F., (2001). Fuzzy control of electro hydraulic steering systems for agricultural vehicles. *Transactions of the ASAE, 44*(6), 1397.
55. Rekow, A. K. W., & Ohlemeyer, H., (2007). Automated Headland Turns the Next Step in Automated Agricultural Machines. In: *Landtechnik*;

Technik Ag Eng. (Vol. 1, pp. 199–209). Hannover. Germany.

56. Romeo, J., Guerrero, J. M., Montalvo, M., Emmi, L., Guijarro, M., Gonzalez-de-Santos, P., & Pajares, G., (2013). Camera sensor arrangement for crop/weed detection accuracy in agronomic images. *Sensors, 13*, 4348–4366.
57. Romeo, J., Pajares, G., Montalvo, M., Guerrero, J. M., Guijarro, M., & Ribeiro, A., (2012). Crop row detection in maize fields inspired on the human visual perception. *The Scientific World Journal, 1*, 1–32.
58. Rovira-Más, F., (2010a). Sensor architecture and task classification for agricultural vehicles and environments. *Sensors, 10*, 11226–11247.
59. Rovira-Más, F., (2010b). General architecture for intelligent agricultural vehicles. In: *Proceedings of Robotics* (Vol. 1, pp. 1–30). Clermont-Ferrand, France.
60. Rovira-Más, F., Zhang, Q., & Hansen, A. C., (2010). Design of intelligent systems. In: *Mechatronics and Intelligent Systems for Off-Road Vehicles* (Vol. 1, pp. 249–269). Springer: London, UK.
61. Rovira-Más, F., Zhang, Q., Reid, J. F., & Will, J. D., (2003). Machine vision-based automated tractor guidance. *Int. J. Smart Eng. Syst. Des., 5*, 467–480.
62. Scaramuzza, D., Achtelik, M. C., Doitsidis, L., Friedrich, F., Kosmatopoulos, E., Martinelli, A., & Gurdan, D., (2014). Vision-controlled micro flying robots: From system design to autonomous navigation and mapping in GPS-denied environments. *IEEE Robotics and Automation Magazine, 21*(3), 26–40.
63. Shrobe, H., Winston, P., Tennenbaum, J., Shaftoe, P., Massaquoi, S., Robertson, P., & Bobrow, R., (2006). CHIP: A cognitive architecture for comprehensive human intelligence and performance. *Electronic Resource, 1*, 177–188. https://pdfs.semanticscholar.org/9992/3ce78ad b0f9b25e16f2d752788b2966ae6a4.pdf.
64. Slaughter, D. C., Giles, D. K., & Downey, D., (2008). Autonomous robotic weed control systems: A review. *Comput. Electron. Agric., 61*, 63–78.
65. Stentz, A., Dima, C., Wellington, C., Herman, H., & Stager, D., (2002). A system for semi-autonomous tractor operations. *Auton. Robot. 13*, 87–104.
66. Suprem, A., Mahalik, N., & Kim, K., (2013). A review on application of technology systems, standards and interfaces for agriculture and

food sector. *Comput. Stand. Interfaces*, *35*, 355–364.

67. Toh, S., Holbrook-Smith, D., Stokes, M. E., Tsuchiya, Y., & McCourt, P., (2014). Detection of parasitic plant suicide germination compounds using a high-throughput Arabidopsis HTL/KAI2 strigolactone perception system. *Chemistry and Biology*, *21*(8), 988–998.
68. Wang, A., Zhang, W., & Wei, X., (2019). A review on weed detection using ground-based machine vision and image processing techniques. *Computers and Electronics in Agriculture*, *158*, 226–240.
69. Wang, G., Do, Y., Cheng, T., Yang, H., Tao, Y., Gu, J., & Yao, L., (2018). Demonstrating printed paper actuator: A low-cost reversible actuation and sensing method for shape changing interfaces. In: *Extended Abstracts of the 2018 CHI Conference on Human Factors in Computing Systems* (Vol. 1, pp. 1–4).
70. Xue, J., Zhang, L., & Grift, T. E., (2012). Variable field-of-view machine vision-based row guidance of an agricultural robot. *Comput. Electron. Agric., 84*, 85–91.
71. Ye, P., Wang, X., Xiong, G., Chen, S., & Wang, F. Y., (2020). TiDEC: A two-layered integrated decision cycle for population evolution. *IEEE Transactions on Cybernetics, 1*, 3–36.
72. Zhang, Q., Reid, J. F., & Noguchi, N., (1999). Agricultural vehicle navigation using multiple guidance sensors. In: *Proceedings of the International Conference on Field and Service Robotics* (Vol. 1., pp. 293–298).
73. Zheng, L., Zhang, J., & Wang, Q., (2009). Mean-shift-based color segmentation of images containing green vegetation. *Comput. Electron. Agric., 65*, 93–98.
74. Zolfagharian, A., Kouzani, A. Z., Khoo, S. Y., Moghadam, A. A. A., Gibson, I., & Kaynak, A., (2016). Evolution of 3D printed soft actuators. *Sensors and Actuators A: Physical, 250*, 258–272.

Chapter 6

Human-Robot Interaction in Agricultural Systems

CONTENTS

6.1. INTRODUCTION

The lack of human labor force is currently experienced in Argentina and Chile for farm tasks, as a sign of numerous features: better living conditions, social adversities, and mining industry (Chile) which offers better and more stable salaries in cities (Argentina). The former issue has direct results in both countries' local economies: Every year, primary farm production is decreased (Traub Ramos, 2013). Automated technology plays a crucial role, in this background since robots can replace human labor force (in some growing's) (Pedersen et al., 2006). However, there should be a gradual replacement. The transformation to fully automated from human dependent agriculture has significances in the farmer, the national (and international), and field workers' regulations (Cheein et al., 2015; Vasconez et al., 2019). Currently, robotic applications are evolving. At present, quick development in personal and service robots is being experienced, providing that robot technology is affecting from factories to personal and service applications (Garrell and Sanfeliu, 2012).

A unique framework is offered by automated agriculture for robotic developments. Although for numerous decades, robotic machinery is existing in the industry (mainly due to the use of robot manipulators in assembling lines), till today due to safety conditions, humans, and robots do not share the workplace (Forlizzi and DiSalvo, 2006). On the other hand, secondary, and primary agricultural tasks can be automated by precision agriculture, such as stated in Pedersen et al. (2006). Advantages and disadvantages of the flexible automation of farms are focused on this work: with field workers, service units sharing the same workplace, and simultaneously sharing the agricultural jobs. For farm robotics, this is a near-future scenario, in recommendations of the authors (Sheridan, 2016; Baxter et al., 2018) (Figure 6.1).

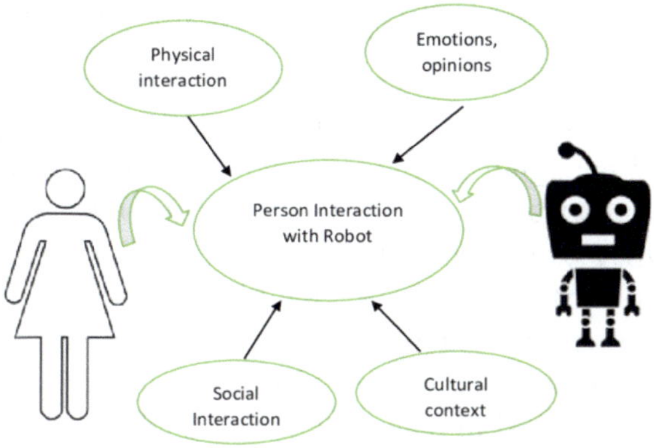

Figure 6.1: *Human-robot interaction mechanism.*

Source: https://www.researchgate.net/figure/General-framework-of-human-robot-interaction_fig1_329057569.

It is to be noted when sharing the same agricultural tasks and the same workplace, the reaction of the field worker to such a situation. Improvement in agricultural production is the primary objective; hence, service units should be capable of interacting with humans regardless of their distrust or probably fear with the machinery (Leahy and Saridis, 1989; Kruse et al., 2013). Moreover, the coupling between robots and field workers should be performed in such a way that humans should feel safe and comfortable in the existence of robots. The scientific literature currently does not cover both these aspects, within an agricultural framework. Within the Human-Robot Interaction (HRI) field, the latter can be circumscribed, i.e., the interaction dynamics between humans and robots (Feil-Seifer and Matarić, 2011). Human-aware robot navigation systems in HRI are dedicated to face problems such as ethics, regulations, socially accepted motions (regarding social rules), safety, acceptance by the workers (including comfort, job redundancy threat, safe perception, among others), natural motions (imitate smoothness of human motions) and human comfort (respecting personal space, direction, and velocity of approach). Flexible automation is presented in this chapter in agriculture: the issue of HRI on the farm environment (Adamides et al., 2013; 2017a, b).

6.2. CONCEPTS OF HUMAN-ROBOT INTERACTIONS

Social robots as part of a heterogeneous group are embodied agents: a society of humans or robots (Dautenhahn and Billard, 1999). They are capable of identifying and engaging in social interactions, they have background information (interprets and perceives the world in terms of their own experience) and they can learn and communicate with each other. In HRI designs, attitude toward robots about what kind of conduct they may show when interacting with robots and the way people perceive robots has resulted to be decisive. T examine human attitudes towards intelligent service robots, a survey has been described in Khan et al. (1998) (Koay et al., 2007).

Additionally, distinguishing the robots that involve *conventional* HRI and interactive robots is important. For domains in which robots must display peer-to-peer communication skills, socially interactive robots are important, either because the main function of the robot is to communicate socially with people or because such skills are essential for solving definite tasks. In the end, *human social* characteristics must be exhibited in a socially interactive robot (Fong et al., 2003).

The most extensively used concepts in the bibliography are remarked in this section that has been applied as a criterion of design regarding HRI, especially with the Human-Aware navigation field.

6.2.1. Social Embeddedness

Usually, social robots are designed as companions, assistants, or pets, as well as the more conventional role of servants. However, in many regions, the absence of human labor force has opened new motivated challenges to develop and introduce sensing technologies and agricultural automation for both secondary tasks (grove supervision, hauling, weed detection, mowing), and primary tasks (seeding, spraying, harvesting, and fertilizing), where humans and robots interact with each other without cooperation by sharing the workplace and working together cooperatively. There are several classes of social robots as presented in Fong et al. (2003) in terms of (1) the intricacy of the communication situation that can be maintained, and (2) how good the robot can upkeep the social model that is attributed to it. It is expected an agricultural service, in this way, that should be socially situated, socially embedded and socially receptive when cooperating with fieldworkers, where:

- Socially receptive. Socially inactive robots but that can advantage from communication (e.g., learning skills by simulated).
- Socially situated. Differentiate between various objects and other social agents in the environment.
- Socially embedded. (a) located in a social environment and cooperate with humans and other agents; (b) physically attached with their social environment; and (c) at least partly conscious of human interactional structures.

6.2.2. Socially-Acceptable Movements

Terms like *naturalness* and *comfort* are used roughly in the bibliography. To improve the social acceptance of such devices, according to cultural norms, robots are expected to behave naturally when interacting. Thus, the taxonomy of socially suitable actions is summarized in Figure 6.2. Briefly, human-aware motions are referred by *acceptable movements,* where like the physic safety, the psychological aspect of human thought is so appropriate, and that the idea of guaranteeing a *sense of security* has been promoted by that inclusive. To imitate and understand the predictable, smooth, and legible human motion, this approach is used in such a manner to encourage the interaction and evade discomfort between humans.

The nonexistence of stress and annoyance for humans is *Human comfort* in interaction with robots. It is not the same as safe: even the motion of the robot is safe, the human may still not feel the motion to be safe. To get a comfortable situation, three aspects are considered essentials: sociability, naturalness, and intentionality.

Sociability is the connection too obvious high-level cultural conventions. Subjects like appreciation of verbal and non-verbal communication, emotion/ gestures, and other perception mechanisms are part of this feature.

Regarding similarity, between humans and robots, the relationship is *Naturalness* in low levels of performance patterns, such as velocities, shapes, and dynamics. It struggles to reproduce nature, such as the motion of humans, as the goal conduct to be restructured by a robot. Though not all features are appropriate for transfer to robots from humans, due to the dissimilar capabilities of robots compared to humans.

Understanding the intentions of a robot intuitively by the person is termed as *Legibility,* or more markedly, is the capability of the robot to act in such a way that human partner can easily infer the intentions of the robot and does not feel afraid or aggravated of the robotic movements (Kruse et al., 2012).

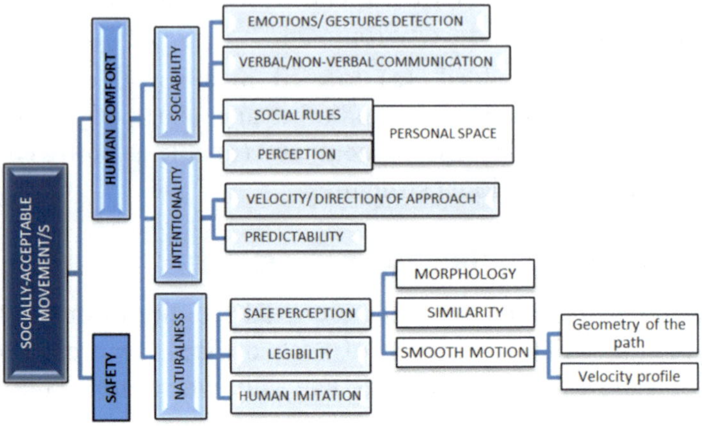

Figure 6.2: Taxonomy of socially acceptable movements.

Source: researchgate.net/publication/277952602_Human-Robot_Interacion_in_Precision_Agriculture_Sharing_the_Workspace_with_Service_Units.

In literature (DiSalvo et al., 2002; McColl and Nejat, 2012), further analysis of human-robot communication can be found.

There is a strong influence of human knowledge, prior exposure, and preconceptions to the robot (or similar robots) on the interaction process. A significant role is played by the characteristics of the robot design (dialogue, embodiment, etc.) within the interaction. Likewise, over time human knowledge will completely shape his judgment, i.e., as a human gains understanding with the robot, early impressions will change. The structure and form of the robot are crucial because it aids develop social expectations. Additionally, the relative strangeness or familiarity of robot morphology can have a significant influence on expressiveness, desirability, and accessibility. The human aptitude to interact with the robot may also be constrained by the choice of a given form. Physical appearance biases interaction, without a doubt.

Hence, a major difference between socially interactive and conventional robots is how a human sees a robot founds expectations that monitor his communication with it (Carton et al., 2012; Ratsamee et al., 2013). This perception, particularly of robot autonomy, capabilities, and intelligence, is affected by various factors, both extrinsic and intrinsic to the person.

6.3. HUMANS, ROBOTS, AND FARMABLE FIELD

In secondary (weed detection, mowing, grove supervision, and hauling) and primary tasks, (fertilizing, harvesting, spraying, seeding) human labor force is present, in precision agriculture. In this context, to share tasks and space with the field workers, service units can be designed. The general architecture of the HRI system is shown in Figure 6.3 applied to agricultural procedures.

Figure 6.3 portrays the most significant cores in such HRI architecture considering the robotics in agriculture can be seen as a flexible automation based system (coexistence of manual tasks and automated): the environment and the machinery, the man (field workers). Following, each stage in Figure 6.3 is described in detail.

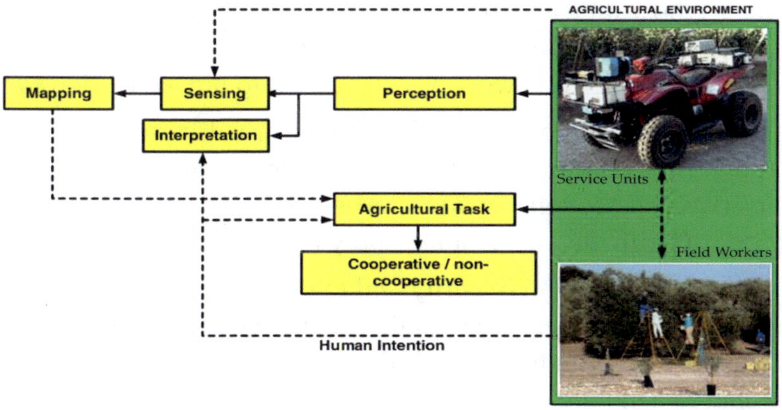

Figure 6.3: The general architecture of an HRI for flexible robotic automation of farms.

Source: researchgate.net/publication/277952602_Human-Robot_Interacion_in_Precision_Agriculture_Sharing_the_Workspace_with_Service_Units.

6.3.1. Service Units

The four designing cores of a service unit as stated by Cheein and Carelli (2013) are sensing, navigation, mapping, and action. Regarding the environment, through its sensors, a service unit, should be able to obtain information and to analyze it to extract valued data (Pölönen et al., 2013).

In this framework, depth cameras, artificial vision systems, range sensors (such as LiDARs) the most extensively used (Rosell et al., 2009; Zarco-Tejada et al., 2014). Their usage is mainly in the extraction of the biological features from features and farms relevant to the vehicle navigation system. The sensors are used to obtain dynamic features along with the static ones, such as other service units and field workers in the environment (Chéné et al., 2012).

To establish a model (i.e., map) of the surrounding environment, the information obtained by the sensors is normally used, to maintain a geographical tracking of information and tasks and to localize the vehicle related to the agricultural activity (Stafford, 2000; AutCheein et al., 2012).

The behavior in which machinery moves within the environment, maneuverability constraints (due to the semi-structured nature of the environment), how it deals with slippage and to fulfill the agricultural task how it plans its motion while interacting with the environment (including other service units and field workers) are related to the navigation system.

Lastly, the agricultural application is concerned with the core *action.* In this framework, as previously stated, a service unit can be used, for secondary (weed detection, mowing, grove supervision, and hauling) and primary tasks (fertilizing, harvesting, spraying, and seeding).

However, to protect the integrity and to optimize the agricultural tasks, the service units should be able to understand the intentions of the workers, when working in environments with field workers. For instance, a service unit is aiding as a mobile deposit, when a field worker is collecting olives in an olive grove, then the machinery should be able to interpret the intentions of the field workers to reach itself to a worker for an appropriate dispose of the olives within the deposit. When the machinery has to evade crash with the field worker is another example: the human motion should be interpreted by the service unit.

6.3.2. Field Workers' Perception

The use of robotized machinery (particularly, autonomous robots) in agricultural tasks, as mentioned previously and interacting with field workers, will certainly influence the perception of the field workers. In such a framework, it is vital to analyze and obtain the response of the human to the flexible automated system. Specifically, the following matters should be carefully considered:

- What are the feelings of the field worker about sharing agricultural tasks with autonomous robots? Which are his/her main apprehensions?
- Does the fieldworker see the machine as a danger to his/her work? The answers are of critical significance since field workers could feel that machines are substituting them. The latter might describe the farmer's agricultural policies and modalities.
- What things the fieldworker would advance in the machinery performance?

In the success of the flexible automation process, the perception of the fieldworker regarding the service units is although simple but vital factor since, as harmony and social rules in every working environment should apply in the advantages of the agricultural production (Salah et al., 2012).

6.3.3. Environment Constraints

Within farmland, the environmental constraints are identified and classified constantly with the natural inconsistency of this partly structured environment, which is open to overcoming all of these constraints and the likelihood of adding structure for robotic applications to simplify the HRI. For intensive agriculture, the fact is especially feasible where pre-existing natural structure (e.g., quasi-symmetric plantation, grove lines) is obvious. *Temporal and spatial variability* are recognized aspects of temporal and spatial variability concentrated on productivity and precision agriculture (Zhang et al., 2002). Various aspects of temporal and spatial variability have been identified by Zhang et al. (2002). However, environmental restrictions have been indirectly defined that must be taken into account within the work of service units and interaction states within farmland as follows:

- **Yield Variability:** Knowledge of present and historical yield distributions will be essential to prioritize scouting tasks of service units and recognize high-presence zones of humans in reaping.
- **Field Variability:** Field topography factors such as slope, proximity to field streams and boundary, elevation, etc., are essential for the design and navigation of traction and slippage control methods of the service units.
- **Soil Variability:** There are two ways in which service units will be affected; first (regarding the task) that could be focused on to sense or manage the certain soil properties; second related to

the navigation because of the inconsistency of the soil density, physical properties-texture, and mechanical strength.

- **Crop Variability:** Crop height, density, biophysical properties, nutrient/water stress are needed to buy the service units in two manners: to avoid damage the crops, planning of navigation, and to determine the health of the plantation and its subsequent treatment, sensing of parameters.
- **Variability in Anomalous Factors:** The continual and precise detection of insects, disease infestation, weed, and nematode as well as hay and wind damage, is critical in the plantation during the care tasks of the service units.
- **Management Variability:** Seeding rate and tillage practice; pesticide and fertilizer application; crop rotation; and irrigation pattern are factors that must be known to service unit to avoid interference in the manual labor during its incursion in the workplace. This kind of variability is both temporal and spatial.
- Mostly, variability's in climate parameters are temporal. However, supporting decision making for fertilizer applications, intensive precipitation monitoring across fields is also significant.

6.3.4. Handling Environment Constraints

There are two approaches by which constraints due to environmental variability can be accomplished: the sensor-based approach and the map-based approach. The second suggests simply navigating optimally and safely through the shared workplace (the map is classically made based on stem positions, in grove plantations), or producing a site-specific map and using this to articulate tasks (e.g., monitoring a variable-rate applicator of pesticides). Other advantages are also presented by the mapping approach: without extra perception systems, the option of ease in making differences between the external objects or individuals and the environment. Contrary to this, real-time sensors are used in the sensor-based approach to cultivate the tasks in an *on-the-go* fashion and inspire the navigation, where a devoted device must be used to detect human individuals.

Because the sensor-based monitoring of field and soil variability is too expensive, unavailable, and not accurate, most experimental precision-agriculture systems are map-based systems. In this manner, to predict the location of humans or to plan trajectories (tasks), maps will aid the development of autonomous navigation.

Moreover, prediction of disease variability and crop growth will be aided by the development of yield potential maps based on anticipated climatic circumstances and establishes an ideal tool for rational farm-work scheduling and agricultural risk assessment.

6.3.5. HRI in Agriculture: Classification

Existing taxonomies are evaluated in this section aimed at characterizing the HRI and for adjusts them for precision agriculture applications. For that reason, the Yanco and Drury (2004) taxonomy is applied.

The main efforts are concentrated on the reaping and sowing stage, according to the tasks done in the agricultural environment. Farm maintenance and fertilization are some other tasks to be estimated with the HRI. There are various variables on which task criticality depends, but criticality is medium under normal security parameters, taking into account that the *bad task* risk is only quantified economically.

Generally, the farm robot morphology is functional today because, to get a human complement, and not a copy, it is researched. Also, to perform specific tasks in times and amounts impossible for animal or human morphology, the machine is needed. *Symbiotic autonomy,* on the other hand, advocates the fact that the robots have cognitive and physical limitations and it asks people for help, supposing the robot is also aware of some of its limitations whenever its capabilities fall short of dealing with a certain situation (Rosenthal et al., 2010).

Farms with heterogeneous robot teams are involved in robot functionality used to perform exact tasks. Without explicit knowledge of the other robots, the robots act and hence there is no interaction between them. However, under the operation of the human or supervision, these act neatly. According to what is specified by Grabowski and Christiansen (2005), this coordination is human-centered and frames within the average spectrum proposed. When the number of robots increases, this category does not scale well, suggesting the need for higher orders of autonomy in-service units.

It pursues an accommodating HRI in which the robot and human share the jobs with shared aims. Depending on the task, in this paradigm, the human and the robot avoid, follow, or approach them, as diagrammed in Figure 6.4. These intentionality's must respect social rules and must be legible that ensure smooth communication.

Present researches are directed at attaining automated farms where humans and machines coexist in total sync (Ellis et al., 1991), although several social and technological barriers are encountered in this major challenge. It is supposed that recognizing, detecting, and following people is one of the needed functionalities for most future robots who will be involved with their human acquaintances (Latif et al., 2009). Usefulness, appropriateness, and efficiency are the factor that determines the successful farm interaction of the robot. To make connections run smoothly, human beings subconsciously familiarize their behaviors to a communication partner.

The robot is anticipated to adapt to its partner, not only the human in HRI (Mitsunaga et al., 2008). Humans tend to subconsciously treat non-personified objects, as Picard et al. (2002) point out, as they would treat other humans (Figure 6.5).

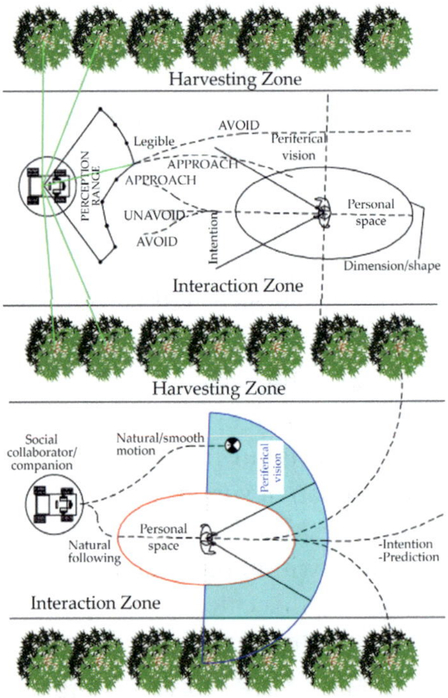

Figure 6.4: Interaction situations in scouting tasks.

Source: https://ieeexplore.ieee.org/document/7125113.

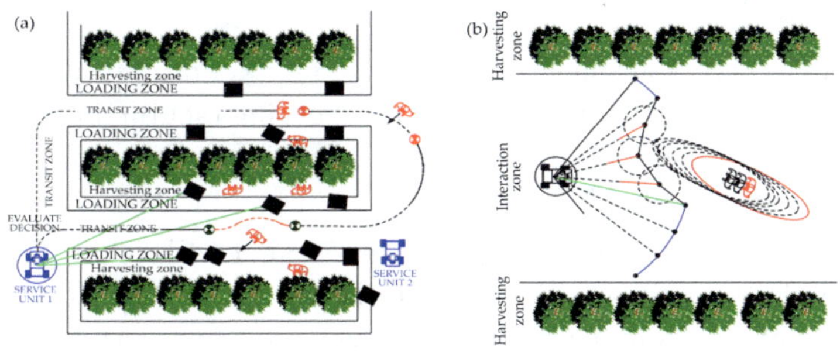

Figure 6.5: HRI in the agricultural harvesting scenario.

Source: https://ieeexplore.ieee.org/document/7125113.

6.3.6. Automated Farms: Pros and Cons

In today's serious agricultural world, farmlands automation has become a competitive benefit. Reduction in the damage to nature, high-quality products, and resource optimization are some of the advantages of the automation in agriculture, becoming a decisive feature in whether or not farmland will continue competitive within the market. There are some negative aspects of automation although it is providing many benefits. Benefits of automated farms are (Herrera et al., 2014):

- Sustainable agriculture;
- High-efficiency;
- Enhancing agricultural input (herbicides and pesticides) and therefore more repeatability, healthy products, and precision;
- Fewer employee costs; and
- In intensive agriculture maintaining yield of traditional practices without scaling the arable land.

Following are the disadvantages of automated farms:

- Less versatility;
- High primary investment;
- Possible social effect on fieldworkers; and
- Specialized personal on initial stages.

It is to be observed that the advantages and disadvantages stated above are essentially connected to the growing and the context (e.g., countries 'rules).

REFERENCES

1. Adamides, G., Katsanos, C., Christou, G., Xenos, M., Kostaras, N., & Hadzilacos, T., (2013). Human-robot interaction in agriculture: Usability evaluation of three input devices for spraying grape clusters. *Innovation*, *24*, 20–27.
2. Adamides, G., Katsanos, C., Constantinou, I., Christou, G., Xenos, M., Hadzilacos, T., & Edan, Y., (2017b). Design and development of a semi-autonomous agricultural vineyard sprayer: Human-robot interaction aspects. *Journal of Field Robotics*, *34*(8), 1407–1426.
3. Adamides, G., Katsanos, C., Parmet, Y., Christou, G., Xenos, M., Hadzilacos, T., & Edan, Y., (2017a). HRI usability evaluation of interaction modes for a teleoperated agricultural robotic sprayer. *Applied Ergonomics*, *62*, 237–246.
4. AutCheein, F. A., De la Cruz, C., Carelli, R., & Bastos, F. T. F., (2011). Autonomous SLAM-based assisted navigation for a robotic wheelchair in restricted environments. *Iberoamerican Journal of Automation and Industrial Informatics RIAI*, *8*(2), 81–92.
5. Baxter, P., Cielniak, G., Hanheide, M., & From, P., (2018). Safe Human-Robot Interaction in Agriculture. In: *Companion of the 2018 ACM/IEEE International Conference on Human-Robot Interaction* (Vol. 1, pp. 59–60).
6. Carton, D., Turnwald, A., Kühnlenz, K., Wollherr, D., & Buss, M., (2012). Proactive human approach in dynamic environments. In: *2012 IEEE/RSJ International Conference on Intelligent Robots and Systems* (Vol. 1, pp. 3320–3321). IEEE.
7. Cheein, F. A. A., & Carelli, R., (2013). Agricultural robotics: Unmanned robotic service units in agricultural tasks. *IEEE Industrial Electronics Magazine*, *7*(3), 48–58.
8. Cheein, F. A., Herrera, D., Gimenez, J., Carelli, R., Torres-Torriti, M., Rosell-Polo, J. R., & Arnó, J., (2015). Human-robot interaction in precision agriculture: Sharing the workspace with service units. In: *2015 IEEE International Conference on Industrial Technology (ICIT)* (Vol. 1, pp. 289–295). IEEE.
9. Chéné, Y., Rousseau, D., Lucidarme, P., Bertheloot, J., Caffier, V., Morel, P., & Chapeau-Blondeau, F., (2012). On the use of depth camera for 3D phenotyping of entire plants. *Computers and Electronics in Agriculture*, *82*, 122–127.

10. Dautenhahn, K., & Billard, A., (1999). Bringing up robots or-the psychology of socially intelligent robots: From theory to implementation. In: *Proceedings of the Third Annual Conference on Autonomous Agents* (Vol. 1, pp. 366–367).
11. DiSalvo, C. F., Gemperle, F., Forlizzi, J., & Kiesler, S., (2002). All robots are not created equal: The design and perception of humanoid robot heads. In: *Proceedings of the 4th Conference on Designing Interactive Systems: Processes, Practices, Methods, and Techniques* (Vol. 1, pp. 321–326).
12. Ellis, C. A., Gibbs, S. J., & Rein, G., (1991). Groupware: Some issues and experiences. *Communications of the ACM, 34*(1), 39–58.
13. Feil-Seifer, D., & Matarić, M., (2011). People-aware navigation for goal-oriented behavior involving a human partner. In: *2011 IEEE International Conference on Development and Learning (ICDL)* (Vol. 2, pp. 1–6). IEEE.
14. Fong, T., Nourbakhsh, I., & Dautenhahn, K., (2003). A survey of socially interactive robots. *Robotics and Autonomous Systems*, *42*(3, 4), 143–166.
15. Forlizzi, J., & DiSalvo, C., (2006). Service robots in the domestic environment: A study of the roomba vacuum in the home. In: *Proceedings of the 1st ACM SIGCHI/SIGART Conference on Human-Robot Interaction* (Vol. 1, pp. 258–265).
16. Garrell, A., & Sanfeliu, A., (2012). Cooperative social robots to accompany groups of people. *The International Journal of Robotics Research, 31*(13), 1675–1701.
17. Grabowski, R., & Christiansen, A., (2005). *A Simplified Taxonomy of Command and Control Structures for Robot Teams* (Vol. 1, pp. 1–21). MITRE CORP MCLEAN VA.
18. Herrera, D., Gimenez, J., Roberti, F., Auat, F., & Carelli, R., (2014). Interacción humano-robot: Navegación socialmente aceptable. *VIII Jomadas Argentinas De Robótica, 1,* 1–32.
19. Khan, Z., (1998). Attitudes towards intelligent service robots. *IPL Technical Report, 1*, 1–22.
20. Koay, K. L., Syrdal, D. S., Walters, M. L., & Dautenhahn, K., (2007). Living with robots: Investigating the habituation effect in participants' preferences during a longitudinal human-robot interaction study. In: *RO-MAN 2007-The 16th IEEE International Symposium on Robot and*

Human Interactive Communication (Vol. 1, pp. 564–569). IEEE.

21. Kruse, T., Basili, P., Glasauer, S., & Kirsch, A., (2012). Legible robot navigation in the proximity of moving humans. In: *2012 IEEE Workshop on Advanced Robotics and its Social Impacts (ARSO)* (Vol. 1, pp. 83–88). IEEE.
22. Kruse, T., Pandey, A. K., Alami, R., & Kirsch, A., (2013). Human-aware robot navigation: A survey. *Robotics and Autonomous Systems, 61*(12), 1726–1743.
23. Latif, H. O., Sherkat, N., & Lotfi, A., (2009). Fusion of automation and teleoperation for person-following with mobile robots. In: *2009 International Conference on Information and Automation* (Vol. 1, pp. 1240–1245). IEEE.
24. Leahy, Jr. M. B., & Saridis, G. N., (1989). Compensation of industrial manipulator dynamics. *The International Journal of Robotics Research, 8*(4), 73–84.
25. McColl, D., & Nejat, G., (2012). Affect detection from body language during social HRI. In: *2012 IEEE RO-MAN: The 21st IEEE International Symposium on Robot and Human Interactive Communication* (Vol. 1, pp. 1013–1018). IEEE.
26. Mitsunaga, N., Smith, C., Kanda, T., Ishiguro, H., & Hagita, N., (2008). Adapting robot behavior for human--robot interaction. *IEEE Transactions on Robotics, 24*(4), 911–916.
27. Pedersen, S. M., Fountas, S., Have, H., & Blackmore, B. S., (2006). Agricultural robots—system analysis and economic feasibility. *Precision Agriculture, 7*(4), 295–308.
28. Picard, R. W., Wexelblat, A., & Clifford, I. N. C. I. N. I., (2002). Future interfaces: Social and emotional. In: *CHI'02 Extended Abstracts on Human Factors in Computing Systems* (Vol. 1, pp. 698–699).
29. Pölönen, I., Saari, H., Kaivosoja, J., Honkavaara, E., & Pesonen, L., (2013). Hyper spectral imaging based biomass and nitrogen content estimations from light-weight UAV. In: *Remote Sensing for Agriculture, Ecosystems, and Hydrology XV* (Vol. 8887, p. 88870J). International Society for Optics and Photonics.
30. Ratsamee, P., Mae, Y., Ohara, K., Kojima, M., & Arai, T., (2013). Social navigation model based on human intention analysis using face orientation. In: *2013 IEEE/RSJ International Conference on Intelligent Robots and Systems* (Vol. 1, pp. 1682–1687). IEEE.

31. Rosell, J. R., Llorens, J., Sanz, R., Arnó, J., Ribes-Dasi, M., Masip, J., & Gil, E., (2009). Obtaining the three-dimensional structure of tree orchards from remote 2D terrestrial LIDAR scanning. *Agricultural and Forest Meteorology, 149*(9), 1505–1515.
32. Rosenthal, S., Biswas, J., & Veloso, M., (2010). An effective personal mobile robot agent through symbiotic human-robot interaction. In: *Proceedings of the 9th International Conference on Autonomous Agents and Multiagent Systems:* (Vol. 1, pp. 915–922). International Foundation for Autonomous Agents and Multiagent Systems.
33. Salah, A. A., Ruiz-del-Solar, J., Mericli, C., & Oudeyer, P. Y., (2012). Human behavior understanding for robotics. In: *International Workshop on Human Behavior Understanding* (Vol. 1, pp. 1–16). Springer, Berlin, Heidelberg.
34. Sheridan, T. B., (2016). Human-robot interaction: Status and challenges. *Human Factors, 58*(4), 525–532.
35. Stafford, J. V., (2000). Implementing precision agriculture in the 21st century. *Journal of Agricultural Engineering Research, 76*(3), 267–275.
36. Traub, R. A., (2013). *First Quarter GDP and Agriculture, 1*, 1–33.
37. Vasconez, J. P., Kantor, G. A., & Cheein, F. A. A., (2019). Human-robot interaction in agriculture: A survey and current challenges. *Biosystems Engineering, 179*, 35–48.
38. Yanco, H. A., & Drury, J., (2004). Classifying human-robot interaction: An updated taxonomy. In: *2004 IEEE International Conference on Systems, Man and Cybernetics (IEEE Cat. No. 04CH37583)* (Vol. 3, pp. 2841–2846). IEEE.
39. Zarco-Tejada, P. J., Diaz-Varela, R., Angileri, V., & Loudjani, P., (2014). Tree height quantification using very high resolution imagery acquired from an unmanned aerial vehicle (UAV) and automatic 3D photo-reconstruction methods. *European Journal of Agronomy, 55*, 89–99.
40. Zhang, N., Wang, M., & Wang, N., (2002). Precision agriculture: A worldwide overview. *Computers and Electronics in Agriculture, 36*(2/3), 113–132.

Chapter 7

Automatic Spraying Robots for Agricultural Use

CONTENTS

7.1. INTRODUCTION

Identification and monitoring of plant diseases, controlled use of pesticides and fertilizers, and controlled irrigation are involved in the responsibility of managing and controlling plant development from the initial stage to the mature harvest stage. The technology of wireless sensors has been explored in this chapter for distant real-time monitoring of vigorous farm parameters like environmental temperature, moisture content of the humidity (Camargoa and Smith, 2009; Al-Hiary et al., 2011). The technique of image processing has also been employed on plant leaves for vision-based automatic disease detection. Design and construction of an autonomous mobile robot have been vigorously described in this chapter featuring spraying mechanism and growth monitoring for fertilizer, pesticide, and water to apply in a plant nursery or agriculture, plant disease detection (Prema et al., 2012; Gutierrez et al., 2014). A portable, compact, and well-founded platform has been provided to realize this work that can examine the growth of the plant and can identify the disease and can survey the farmland automatically and accordingly spray fertilizer, pesticide, and water to the plant. By providing real time information about the environment and plants using fundamental principles of Image processing and Sensor technology, the Internet, this method will help farmers make the right decisions (Zhang et al., 2002; Rawidean et al., 2014).

The economy of the nation is decided by agriculture and the main business of Indian people is agriculture. For plant growth, the essential nutrients are normally produced in its surroundings. Conditions of the environment are a decisive factor in the plant development process where a plant grows (Pydipati et al., 2006; Jian-Sheng, 2014) (Figure 7.1).

Figure 7.1: Automatic spray painting robot.

Source: http://www.slcoating.com/product/automatic-spray-painting-robot.html.

In the environment, there are necessary parameters like ambient temperature, moisture, humidity, CO_2, light, etc. The farmer can get help and get much acquainted with any of the potential difficulties that will disturb the health of the plants by a deep understanding of all these influences and their relationships and thereby more suitable and precise measures can be taken to get rid of these difficulties.

The plant development process is most affected by the factor of temperature. A disparate temperature range is presented by each plant species within which they can develop normally. The processes required for the plant's life stop above this range as the enzymes become inactive. Humidity controls the loss of moisture from the plant. The development process of the plant will get affected as humidity increases because air will also become saturated with vapor which will then limit transpiration and fungal diseases will spread quickly. Other main biological parameters such as disease, pests, and soil also affect the productivity of the crop. To determine the right amount of water to supply and the right time to irrigate, soil moisture is the key parameter that can be used. For improvising the production of a crop, these parameters can be organized by human beings (Junxiong et al., 2009). Many operations from humans are increasingly taken over by the robots where precision in work and repeatability in a routine task are required and also in conditions where human operators are exposed to risk (Snehal et al., 2015). Raising of crops in the greenhouse is one of the tasks involved, where most of the processes on the crop are still achieved manually by the human operators even though they are often extremely continuous (Pilli et al., 2015). To mechanize the irrigation systems, the water content info of soil can be used by using dielectric moisture sensors, and rather than having a static irrigation plan with precise duration, these consequently control the actuators for water. As part of Precision Agriculture, autonomous mobile robots are now relegated for optimizing field management tasks of agriculture to improve crop perception, environment supervision, and economics (Liu et al., 2011).

The technology of wireless sensors has been explored in this chapter for remote real-time monitoring of vigorous farm parameters like moisture content, humidity, and environmental temperature of the soil. The method of image processing has also been employed for vision-based automatic detection of diseases on plant leaves. A portable, compact, and well-founded platform is provided to realize this that can automatically survey the farmland and can examine the growth of the plant and also can recognize the disease and accordingly spray fertilizer, pesticides, and water to the plant.

This method will aid farmers in making the right decisions (Belforte et al., 2006; Oberti et al., 2016).

The conditions of the environment are the basis of the plant development process, where the plant grows. The essential parameters like ambient temperature, humidity, moisture, CO_2, and light, etc are included in the environment. Deep understanding of these parameters and their relationships can aid the farmers to get familiar with any of the potential difficulties that will impact the health of the plants and thereby more accurate and appropriate measures can be taken to eliminate these issues.

7.2. BACKGROUND OF AUTOMATIC AGRICULTURAL SPRAYING

In the agriculture field, the discovery of plant diseases and their extent, for the crop quality has always been the main concern. Many researchers have been working on this problem area to date. Many disease detection systems proposed earlier based on image processing techniques. Below are some already established systems area: to examine the type and extend of diseases, the sprayer mechanism that sprays the obligatory pesticide, Pilli et al. (2015) has given AI-based embedded algorithms that perform color transformations by picking the region of interest (ROI). A measurement technique for pest recognition on leaves has been provided by Chenelli et al. (2015). By using a binocular stereo vision system, this system allows automatic spraying of pesticide to the pest's position on the leaves. Color co-occurrence method (CCM) has been in research by R. Pydipati et al. (2006) to check whether texture-based intensity, saturation, and hue color features together with numerical organization algorithms could be used to classify normal and diseased citrus leaves under laboratory conditions. The processing algorithm has been shown by A Camaro and Smith (2009) that includes a change of the RGB image of the diseased leaf of the plant into another color transformations I3b, I3a, and H.

For spraying pesticides in a greenhouse environment, a Navigation system has been offered by Liu et al. (2011) that mainly includes electromagnetic sensors and controllers. To guarantee that the optimum quantity of spray and evading of human exposure to the hazardous chemicals, an autonomous pesticide spraying device would be of the greatest significance. In the same context, for precise use of water, irrigation systems are also required, Joaqun Gutirrez et al. (2014) has given solution to this problem.

The Block diagram of the proposed system has been displayed in Figure 7.2. An important element of the system has been shown in each block. ARM7 is used as the control unit, as can be seen. For environment monitoring, numerous sensors also have combined. For spraying purposes, three water motors are included of each liquid (Tillett, 1991; Mulla and Khosla, 2016). The vision capability of the system is enabled by the webcam. For wireless communication purpose, GSM, and Zigbee are used.

7.3. DESIGN CONSIDERATIONS OF THE SYSTEM

Without a hardware and software combination, any system design is incomplete in the current condition. The proposed system in this section is explained in a further two sections (Moltó et al., 2001; Sammons et al., 2005).

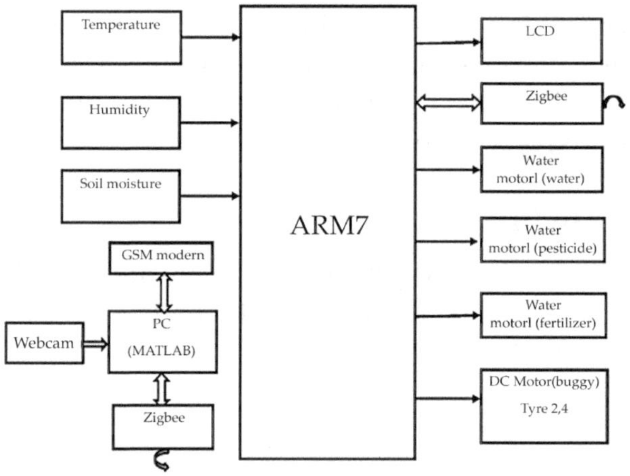

Figure 7.2: Block diagram of the proposed system.

Source: https://www.semanticscholar.org/paper/Automatic-Agriculture-Spraying-Robot-with-Smart-Sharma-Borse/e78047d81720414284d08bad-8c78bd98571d4fd4.

7.3.1. Hardware Details

Three sensors (a soil moisture sensor, temperature sensor, and humidity sensor, a microcontroller ARM7, and three containers of 500 ml fir spraying of fertilizer, pesticide, and water are included in the hardware part. To physically run the robotic buggy, two dc motors are required. To capture the images of plants, a webcam is used in the WI-FI environment, through MATLAB Command. For sending SMS to personal computers (PCs) and farmers with MATLAB software, the system also has a GSM modem (Blackmore et al., 2008; Lan et al., 2017).

7.3.2. System Control Section

LPC2138 microcontroller is the system's controlling part as the core controller. The LPC2138 microcontrollers are based on a 32/16 bit ARM7TDMI-S CPU with real-time embedded trace support and emulation. It has 64 pins (Keicher and Seufert, 2000; Javed et al., 2017).

7.3.3. Wireless Sensor Unit

A microcontroller, RF transceiver, power sources, and sensors are included in the sensor unit. Microcontroller LPC2138 is the basis of the wireless sensor unit that controls processes information from Temperature sensor LM35, the soil moisture sensor IC LM358 and humidity sensor (Precision Centigrade Sensors) and radio modem ZigBee.

7.3.4. Driver Modules

To run the application physically DC motors as per the necessities configured in software. The DC motor functions on 12 v. A DC motor driver called L293D is required to drive a DC motor.

7.3.5. Trolley Modules

The entire hardware system and the liquid tanks are carried by the trolley module that is a carrier trolley. The design of an adjustable chassis is the main challenge that could carry a load of 5–10 kgs. Thus, the chassis of the vehicle is made up of wooden and it can step in the variability of composite cement littlie mud, road pavement, grass, gravel, etc. With higher power batteries and big tires, paving is not that necessary it can move over fields and grass too but care should be taken about any pit and pothole (Stafford and Miller, 1993; Itoh et al., 2014).

7.3.6. Spray Modules

Spray head, servos, relays, DC machine, and pumps are included in the spray module. There are three tanks in the spray system. An on-board microprocessor controls the valves electronically. The pump is turned on and off, as per the plant's distance, as the robot passes over reflective places on the ground to enable selective spraying of the diseased. The microcontroller I/O port can control the working status of the pump, to spray from the spray head (Thompson et al., 1991; Fukao et al., 2011).

7.4. SOFTWARE DETAILS

The image processing technique has been used to build the growth and disease identification ability of the ROBOT. Acquisition of an image, processing of image and then segmentation, statistical analysis, feature extraction, and then classification MATLAB software is the basic steps involved in this method used for growth measurement and disease detection based on several algorithms that arrange the software part of the system. Two dissimilar flows are included in this section, one for finding out the growth of the plant and one for the detection of diseases as shown in Figure 7.4. For that reason, on the robot buggy, a wireless camera is mounted and then this camera is connected to PC which has MATLAB software through Wi-Fi connectivity (Hauwiller and Jin, 2001; Nikkilä et al., 2010) (Figure 7.3).

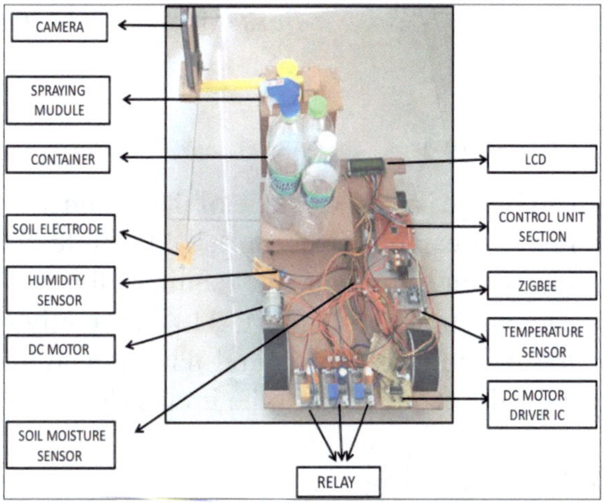

Figure 7.3: Agriculture robot hardware assembly.

Source: https://www.semanticscholar.org/paper/Automatic-Agriculture-Spraying-Robot-with-Smart-Sharma-Borse/e78047d81720414284d08bad-8c78bd98571d4fd4.

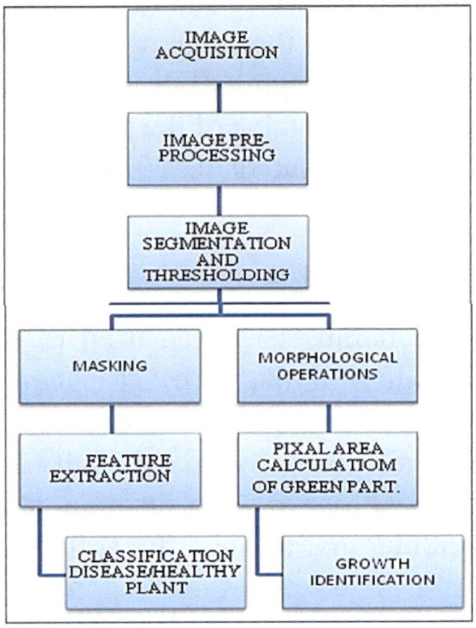

Figure 7.4: The flowchart describes the flow of the System for disease detection and growth identification.

Source: https://www.semanticscholar.org/paper/Automatic-Agriculture-Spraying-Robot-with-Smart-Sharma-Borse/e78047d81720414284d08bad-8c78bd98571d4fd4.

7.4.1. Image Acquisition

The acquisition of an image is the first stage of the vision system. An android mobile phone camera, a camera (webcam), and wireless fidelity (Wi-Fi) internet connection are the components required for the image capture module. An image can be transferred to MATLAB directly from the mobile with the help of the Android-based OS and its applications. Wi-Fi and IP webcam is needed for the acquisition of the image (Franz, 1990).

7.4.2. Image Pre-Processing

Conversion to the HSI image from the RGB image is done in this project. By manipulation of the brightness values, HSI color space is the best for traditional image processing functions, as I consistently dependents on G, R, and B. A greater dynamic range of saturation is provided by the HIS color and thus it is used for saturation and hue manipulation (to regulate the quantity of color). S specifies saturation, H signifies hue that specifies the color purity and I express the light intensity (Hu et al., 2011).

Color space selection for image processing is an essential step. For growth detection in this project, the Hue, Saturation, Value (HSV) color space has been nominated. Frequently used color space is RGB color space and there is a high association between G, B, and R. Unsatisfactory results would be generated by using these values directly. There is a need to transform RGB color space to HSV, to diminish the correlation between the R, G, and B.

The conversion formula to HSV from RGB color space.

$$Hue\ (H) = \begin{cases} 2 - A.Cos\frac{[(R-G)+(R-B)}{\sqrt[2]{(R-G)^2+(R-G)(G-B)}}, & B > G \\ A.Cos\frac{[(R-G)+(R-B)}{\sqrt[2]{(R-G)^2+(R-G)(G-B)}}, & B \leq G \end{cases} \quad (1)$$

$$Saturation(S) = \frac{\max(R,G,B) - \min(R,G,B)}{max(R+G+B)} \quad (2)$$

$$Value(V) = \frac{R+G+B}{255} \quad (3)$$

7.4.3. Image Segmentation

Image segmentation is used to discriminate the various regions in an image that do not overlap with each other (Weizheng et al., 2008).

Image segmentation is done for the detection of disease using K-means clustering. The K-means clustering algorithm efforts to categorize objects into K number of glasses rested on a set of features. By diminishing the sum of squares of distances between the corresponding class or cluster centroid and the objects, the classification is achieved. In this project, three clusters proved to be sufficient.

The masking of the green pixels is done after that on the boundaries. There are two steps involved in this: Most of the pixels of green color are identified at first subsequently to attain a variable threshold value Global image threshold has been used. If the green component in the pixel intensities is less than the projected threshold value then this pixel's G, B, and R components are disappeared. Deleting of pixels with zero components and those at the boundaries of the diseased clusters is the second required step (Esau et al., 2014).

Dissimilar segmentation methods have been used to find the growth of the plant to determine the lesion and the total leaf area pixels. Converting an input image into a grayscale image is involved in this and the large difference in gray values of object and background would be produced by taking the image of the diseased plant with white background. After the region, a filling is done and holes are eliminated in the white region to have the binary image that comprises of the lead region. Lastly, the image is scanned from left to right and top to bottom to count pixels of the total leaf area. The growth of the plant will be resulted as normal if the calculated pixel count of the green part is greater than a specific set point (pixel count) otherwise abnormal.

7.4.4. Computation of Integrated Feature Vector

The Gray Level Co-Matrix (GLCM) is used in this project to extract the statistical data from the image texture.

A gray-level co-occurrence matrix (GLCM) is produced by glam = gray matrix (I) from the input image. By calculation of how repeatedly a pixel with gray-level (grayscale intensity) value I happen horizontally together to a pixel with the value j, GLCM is created by Graycomatrix. Using the 'offsets' parameter, other pixel spatial relationships can also be specified. Calculation of how recurrently a pixel with gray-level (grayscale intensity) value I happen horizontally together to a pixel with the value j is specified by each of the elements (i, j).

Various statistical data is computed by the gray level co-matrix some of them are:

1. Entropy (E):

$$E = -\sum_{i,j} p(i,j) \log(p(i,j)) \quad (4)$$

Entropy indicates the disorder in a system. The system would present a high entropy in case of random distribution. There would be 0 value for entropy measures in an image with a solid tone. The entropy of the smooth and heavy texture is described by the features. The spatial disorder is computed by the texture analysis.

2. Correlation (r):

$$r = \frac{\sum_{i,j}(ij)\mathrm{p(i,j)} - \mu_x\mu_y}{\sigma_x\sigma_y} \quad (5)$$

where p_y and p_x are fractional probability density functions. σ_x σ_y are the Standard deviations of p_x and p_y and $\mu_x\mu_y$ are the means.

3. Energy (e):

$$e = \sum_{i,j} p(i,j)^2 \qquad (6)$$

Energy is precisely contrary to entropy measurement. It is the approximation of homogeneity. Texture with greater homogeneity would outcome in higher energy.

4. Homogeneity (h):

$$h = \sum_{i.j} \frac{1}{1-(i-j)^2} p(i,j) \qquad (7)$$

In the gray level co-matrix, consistency approximation of non-zero entries signifies the homogeneity. If there are a lot of pixels with similar grey levels, GLCM homogeneity would be high and GLCM homogeneity is low when there are varying gray level values. An image having homogeneity value is 1, no variation, and an image having little variations would have a high homogeneity value.

7.4.5. Support Vector Machine (SVM)

For classification, one of the machine learning techniques, Support Vector Machine (SVM) is used. It is a classifier based kernel. Initially, for linear separation, it was developed which was capable of classifying the data in two categories. For identifying healthy leaf and diseased leaf, SVM has been used in our project. For that reason, the data set has been prepared manually. From mobile, the plant's leaf samples were obtained and these leaves are categorized into diseased and normal subsets. Brown and yellow Spots are two types of diseases contained in the samples of the disease subset.

REFERENCES

1. Al-Hiary, H., Bani-Ahmad, S., Reyalat, M., Braik, M., & AL Rahamneh, Z., (2011). Fast and accurate detection and classification of plant diseases. *International Journal of Computer Applications, 17*, 1–43. doi: 10.5120/2183-2754.
2. Belforte, G., Deboli, R., Gay, P., Piccarolo, P., & Aimonino, D. R., (2006). Robot design and testing for greenhouse applications. *Biosystems Engineering*, *95*(3), 309–321.
3. Berenstein, R., & Edan, Y., (2017). Automatic adjustable spraying device for site-specific agricultural application. *IEEE Transactions on Automation Science and Engineering*, *15*(2), 641–650.
4. Blackmore, B. S., Fountas, S., Gemtos, T. A., & Griepentrog, H. W., (2008). A specification for an autonomous crop production mechanization system. In: *International Symposium on Application of Precision Agriculture for Fruits and Vegetables* (Vol. 824, pp. 201–216).
5. Camaro, J. S., & Smith, (2009). An image-processing based algorithm to automatically identify plant disease visual symptoms. *Elsevier Journal of Computer and Electronics in Agriculture* (Vol. 1, pp. 1–19). doi: 10.1016/j.biosystemseng.2008.09.030.
6. Chunlei, V. L., & Xia, J. L., (2015). Vision-based pest detection and automatic spray of greenhouse plant. *International Conference on (ISIE)* (Vol. 1, pp. 10–22). Korea doi: 10.1109/ISIE.2009.5218251.
7. Esau, T. J., Zaman, Q. U., Chang, Y. K., Schumann, A. W., Percival, D. C., & Farooque, A. A., (2014). Spot-application of fungicide for wild blueberry using an automated prototype variable rate sprayer. *Precision Agriculture*, *15*(2), 147–161.
8. Franz, E., (1990). Progress towards a machine vision system for spray-deposit analysis: Software development. *Paper-American Society of Agricultural Engineers*, *90*, 7500–7515).
9. Fukao, N., Kyung, K. H., Fujimoto, K., & Shiratori, S., (2011). Automatic spray-LbL machine based on in-situ QCM monitoring. *Macromolecules*, *44*(8), 2964–2969.
10. Gutierrez, J., Villa-Medina, J. N., Nieto-Garibay, A., & Porta-Gandara, M. J., (2014). Automated irrigation system using a wireless sensor network and GPRS module. *IEEE Transactions on Instrumentation and Measurement* (Vol. 63, pp. 1–26). doi: 10.1109/TIM.2013.2276487.

11. Hauwiller, J. I., & Jin, Y., (2001). U.S. Patent No. 6,236,907, pp. 1–20. Washington, DC: U.S. Patent and Trademark Office.
12. Hu, K., Fu, Z., Ji. R., Wang, J., & Qi, L., (2011). Design and development of variable rate spraying system based on canopy volume measurement. In: *International Conference on Computer and Computing Technologies in Agriculture* (Vol. 1, pp. 402–413). Springer, Berlin, Heidelberg.
13. Itoh, H., Navarro, R., Takeshita, K., Tago, K., Hayatsu, M., Hori, T., & Kikuchi, Y., (2014). Bacterial population succession and adaptation affected by insecticide application and soil spraying history. *Frontiers in Microbiology*, *5*, 457.
14. Javed, M. H., Noor, M. H., Khan, B. Y., Noor, N., & Arshad, T., (2017). K-means based automatic pests detection and classification for pesticides spraying. *International Journal of Advanced Computer Science and Applications*, *8*(11), 236–240.
15. Jian-Sheng, P., (2014). An intelligent robot system for spraying pesticides. *The Open Electrical and Electronic Engineering Journal, 8,* 435–444.
16. Junxiong, Z., Cao, Z., & Geng, C., (2009). Research on precision target spray robot in greenhouse. *Transactions of the Chinese Society of Agricultural Engineering, 25*, 70–73.
17. Keicher, R., & Seufert, H., (2000). Automatic guidance for agricultural vehicles in Europe. *Computers and Electronics in Agriculture*, *25*(1, 2), 169–194.
18. Lan, Y., Shengde, C., & Fritz, B. K., (2017). Current status and future trends of precision agricultural aviation technologies. *International Journal of Agricultural and Biological Engineering*, *10*(3), 1–17.
19. Liu, T., Bin, Z., & Jixing, J., (2011). Electromagnetic navigation system design of the green house spraying robot. *IEEE Second International Conference (MACE)* (Vol. 1, pp. 1–22). Hohhot. doi: 10.1109/MACE.2011.5987400.
20. Moltó, E., Martín, B., & Gutiérrez, A., (2001). Pm—Power and machinery: Pesticide loss reduction by automatic adaptation of spraying on globular trees. *Journal of Agricultural Engineering Research*, *78*(1), 35–41.

21. Mulla, D., & Khosla, R., (2016). Historical evolution and recent advances in precision farming. *Soil-Specific Farming Precision Agriculture, 1*, 1–35.
22. Nikkilä, R., Seilonen, I., & Koskinen, K., (2010). Software architecture for farm management information systems in precision agriculture. *Computers and Electronics in Agriculture, 70*(2), 328–336.
23. Oberti, R., Marchi, M., Tirelli, P., Calcante, A., Iriti, M., Tona, E., & Ulbrich, H., (2016). Selective spraying of grapevines for disease control using a modular agricultural robot. *Biosystems Engineering, 146*, 203–215.
24. Pedersen, S. M., Fountas, S., Have, H., & Blackmore, B. S., (2006). Agricultural robots—system analysis and economic feasibility. *Precision Agriculture, 7*(4), 295–308.
25. Pilli, S. K., Bharathiraja, N., Smith, J. G., & Vivek, D., (2015). eAGROBOT-A robot for early crop disease detection using image processing. *IEEE Sponsored 2nd International Conference on Electronics and Communication System (ICECS), 1,* 1–32. doi: 10.1109/ECS.2014.7090754.
26. Praveena, R., & Srimeena, R., (2015). Agricultural robot for automatic ploughing and seeding. In: *2015 IEEE Technological Innovation in ICT for Agriculture and Rural Development (TIAR)* (Vol. 1, pp. 17–23). IEEE.
27. Prema, K., Senthil, K. N., Dash, S. S., & Sudhakar, C., (2012). Online control of remote operated agricultural robot using fuzzy controller and virtual instrumentation. *IEEE-International Conference on Advances in Engineering, Science, and Management* (IAESM) (Vol. 1, pp. 196–201).
28. Pydipati, R., Burks, T. F., & Lee, W. S., (2006). Identification of citrus disease using color texture features and discriminant analysis. *Computers and Electronics in Agriculture* (Vol. 1, pp. 1–28). doi: 10.1016/j.compag.2006.01.004.
29. Rawidean, M., Mohd, K., Ibrahim, M., & Ahmad, N. H., (2014). Wireless sensor network in precision agriculture application. *IEEE Conference MIMOS* (Vol. 1, pp. 1–24). Malaysia. doi: 10.1109/CITS.2014.6878963.
30. Sammons, P. J., Furukawa, T., & Bulgin, A., (2005). Autonomous pesticide spraying robot for use in a greenhouse. In: *Australian Conference on Robotics and Automation* (Vol. 1(9), pp. 1–23).

31. Snehal, M. D., & Dr. Gengaje, S. R., (2015). ARM-based pesticide spraying robot. *International Journal of Engineering Research and General Science, 3*, 1–30. ISSN: 2091-2730.

32. Stafford, J. V., & Miller, P. C. H., (1993). Spatially selective application of herbicide to cereal crops. *Computers and Electronics in Agriculture, 9*(3), 217–229.

33. Swain, K. C., Nørremark, M., Bochtis, D., Olsen, H. J., Sørensen, C. G., Green, O., & Hameed, I. A., (2009). Automated blob spraying system for agricultural robots. In: *Proceedings of the 10th International Agricultural Engineering Conference* (Vol. 1, pp. 1–33). Bangkok, Thailand. Role of agricultural engineering in advent of changing global landscape. Asian Association for Agricultural Engineering.

34. Thompson, J. F., Stafford, J. V., & Miller, P. C. H., (1991). Potential for automatic weed detection and selective herbicide application. *Crop Protection, 10*(4), 254–259.

35. Tillett, N. D., (1991). Automatic guidance sensors for agricultural field machines: A review. *Journal of Agricultural Engineering Research, 50*, 167–187.

36. Weizheng, S., Yachun, W., Zhanliang, C., & Hongda, (2008). Grading method of leaf spot disease based on image processing. *International Conference on Computer Science and Software Engineering, 1*, 12–14. doi: 10.1109/CSSE.2008.1649.

37. Zhang, N., Maohua, W., & Ning, W., (2002). Precision agriculture-a worldwide overview. *Computers and Electronics in Agriculture Elsevier, 36*, 113–132. doi: 10.1016/S0168-1699(02)00096-0.

Chapter 8 Economics and Applications of Agricultural Robots

CONTENTS

8.1. INTRODUCTION

In warehouses and for industrial production robotic systems have been broadly used for many years, where a controlled environment can be guaranteed. In forestry and agriculture, there has been a vision in the early 1960s about research into driverless vehicles with basic research on projects on autonomous tractors and automatic steered systems (Wilson, 2000; Griepentrog et al., 2006).

An increased interest has been observed recently in agriculture in the development of robotic systems, which has led many specialists to discover the possibilities based on a behavioral approach to cultivate more adaptable and rational vehicles. Combined use of new communication technologies, geographical information systems (GIS), sensor systems, and positioning systems (GPS) have assisted investigators to develop new autonomous vehicles for high-value crops in the horticulture and agriculture sector, along with landscape management (Nørremark and Griepentrog, 2004).

For horticultural and orchards crops, various autonomous prototypes have been described, such as strawberries (Kondo et al., 2005), oranges (Hannan and Burks, 2004), and tomatoes (Chi and Ling, 2004). Moreover, for high-value crops, automated systems based on real-time climatic situations for site-specific irrigation have been defined (e.g., Miranda et al, 2005) (Figure 8.1).

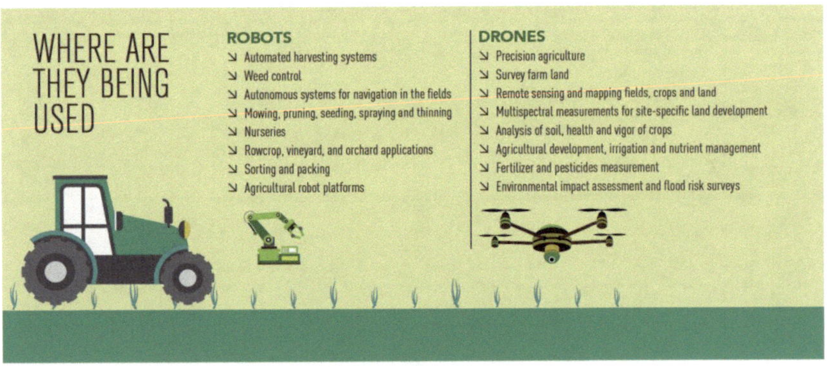

Figure 8.1: Different uses of robots and drones.

Source: https://www.geospatialworld.net/article/drones-and-robots-future-agriculture/.

There are several systems, for field crops autonomous Christmas tree weeder (Have et al., 2005), the Demeter system equipped with GPS and camera for navigation for automated harvesting (Pilarski et al., 2002) and patch spraying the API platform (Bak and Jacobsen, 2004). Moreover, there are automated systems in indoor environments for animal products such as automated cleaning and feeding.

Autonomous and robotic systems are more difficult to develop in the outdoor and open environment, which will be the focus of discussion here, mainly because of safety concerns. The robot's safety system would have to be consistent enough for it to function unattended and autonomously. If the automobile has to be independent, it is comparatively expensive to develop safety systems. In opinion, if a robot has to be attended then the time is restricted by the person otherwise, they can work 224 hours a day. In this regard, depending on the task to be performed, various degrees of autonomy and scenarios have been investigated. Replacement of traditional large tractors by small autonomous machines have been initiated to investigate which would be would be more efficient. All year-round, these vehicles should be able to perform valuable tasks, unattended, and over long periods capable to perform reasonably in a semi-natural environment. Replacing the over-application of fertilizers and chemicals, the small vehicles may also have less ecological influence, with better control harmonized to requirements, needing lower usage of energy and also due to lighter weight causing less soil compaction (Graglia, 2004; Søgaard et al., 2006) (Figure 8.2).

Figure 8.2: MF-Scamp robots for scouting, weeding, and harvesting.

Source:https://www.researchgate.net/publication/221787093_Agricultural_Robots_Applications_and_Economic_Perspectives.

8.2. ECONOMICS

So far, there are very few studies published on the consequences of economics to achieve more maintainable production systems by introducing autonomous field machinery. Conventional vehicles have been compared with autonomous vehicles by Goense (2003), armed with apparatuses having working widths to 120 cm from 50 cm. He showed that it would be more feasible with slight rises in labor costs or with minor decreases in prices of navigation systems if autonomous vehicles can be utilized 23 hours a day. Various other changes have also been discussed by Goense that will disturb the final result such as the part of labor time required out of the total machine tracking system and the total machine time, and there is no need for operators' rest allowance as it gives better utilization of machine working width. Contrary to this, in the form of higher costs in traveling distances, there may be negative effects on service personal (Ali 2004; Søgaard and Lund, 2005).

Moreover, the effects of automation on crop establishment costs for soil tillage and machinery sizes have been analyzed by Have (2004). He estimated that the ratios between a manned tractor and an autonomous tractor in terms of daily working hours, labor requirements, and the price would be 2, 0.2, and 1.2 times respectively. The analysis showed that the shift to automatic control would reduce the implement sizes, tractor size and implement investments to about half which included all timeliness costs and direct machinery costs; as well as it would decrease the sum of annual machinery and tractor costs to approximately 65% and the tractor investment to about 60%. A comparison has been shown by Pedersen et al. (2006 and 2007) between conventional systems and 3 various applications of robot systems for landscape treatment and crop production. In the present chapter, the economic feasibility study depends on some of these descriptions and scenarios (Lund and Sogaard, 2004).

8.3. CROP ESTABLISHMENT

In the light of developing smarter machines, various ideas of crop cultivation and production are now being reconsidered and revisited.

To create a suitable seedbed, it is common that the whole topsoil of a field is upturned with a plow for seedbed preparation and traditional crop establishment. This is a well-known procedure that uses a lot of energy but suits many circumstances. Other approaches with abridged tillage may become an option if this method is turned around to consider the actual seed necessities. To allow uptake of nutrients and water, the seed needs contact with soil moisture; it needs a structure that can allow shoots to grow and roots to develop and that can hold the plant upright. Additionally, space is required with decreased competition from other plant seeds. The soil does not need to be disturbed if the same seed atmosphere can be attained by mixing only the soil within a few centimeters of the actual seed as it can be well-conditioned by natural soil fauna and flora.

Growing of crops in rows is another conventional concept. The only explanation as to why this is done as it would seem is that it needs the simplest kinds of machines. Along each row, seeds are positioned relatively densely. In opinion, the issue is that each plant needs equal access to water, light, nutrients, and air, which are frequently spatially related. By giving seed distribution and more even or equal spacing with precise placement of seeds in a more uniform pattern, intra crop competition can be reduced (Figure 8.3). The main crop is illustrated in Figure 8.3 with weeds of red spots and green spots. The crop is seeded in traditional rows, on the left of Figure 8.3. Here, is possible for the weed to overtake the main crop and grow fast at an early development stage. The crop is dispersed consistently with seed placement, on the right side. In opinion, before the weed develops the main crop will cover the weed and hence decrease the weed pressure. It may be possible to decrease the weed biomass by up to 60% as indicated by the initial trials by the use of uniform patterns with precise seed placement (Weiner and Olsen 2007).

If the position of each developed crop plant is assessed and the location of each seed is known, by its spatial location it will be probable to classify each plant. Improved decision making and management is allowed by the improved information about plant features and allows several more targeted and improved operations that can advance the overall efficiency of crop growth. Various methods could be used as only a small portion of soil is needed to be cultivated. Instead of just one dimension or water-jetting, rotary mechanical tillage in two dimensions on a horizontal or vertical axis (as displayed in the seed map in Figure 8.3) could be used or to create a micro-climate for the seed the injection of a hygroscopic polymer gel could be used.

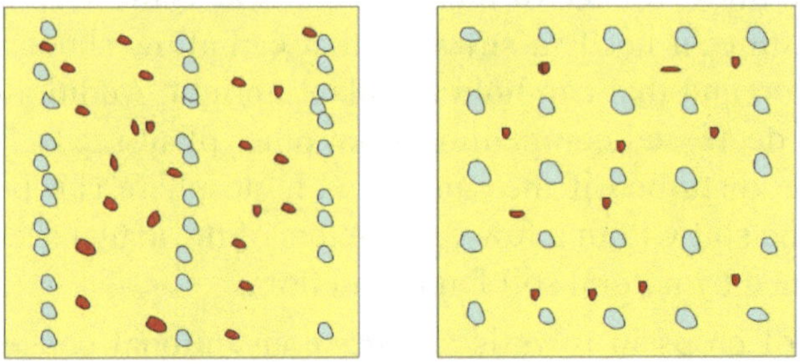

Figure 8.3: Precise seed placement in uniform patterns (Source. Based on Weiner J and Olsen J, 2007. Momentum 2007).

Source: https://www.researchgate.net/publication/221787093_Agricultural_Robots_Applications_and_Economic_Perspectives.

Seed mapping is comparatively simple in practice as infrared sensors mounted below the seed chute and a tilt attuned RTK-GPS could be tailored to the high precision seeder. As the seed drops, it triggers a data logger that records the designated orientation and position of the planter and it cuts the infrared beam. The actual seed position can be calculated by the simple kinematic model (Griepentrog et al., 2005a). Though, for a moving machine, it is difficult to place ultra-high precision of seed. Even there are problems in the most sophisticated commercial machines with reliably unraveling individual seeds and as they pass through the metering mechanisms, they will often damage the seeds. To stop the seed bouncing after impact with the soil it is important to install some mechanisms to ensure that each seed has zero ground velocity (Griepentrog et al., 2005b).

8.4. CROP SCOUTING

The ability to collect accurate and timely information about crop development is an important part of good management. Sampling costs can rapidly outweigh the welfares of spatially adjustable management and quantified data has tended to be expensive (Godwin et al., 2003).

If an automated system could remain within the crop canopy, data collection would be timelier and less expensive for continual monitoring that can be used for evaluating crop status. This could be attained by either placing more costly sensors onto a moving platform or embedding cheap wireless sensors at strategic positions within the crop.

Crop features would include crop height (giving growth rates), senescence, biomass, Leaf Area Index, growth stage, etc. In the early stages, it is difficult to assess crop nutrient status and it becomes very complicated to assess it independently from moisture stress (Christensen et al., 2005).

To monitor growing crops for disease attack and pest, a whole new set of opportunities will become available with the advent of biosensors (Tothill, 2001). Real-time alerts could be sent to the manager as the robotic/ autonomous vehicle could watch the fields constantly viewing for weeds and other threats whenever certain circumstances were faced. By monitoring environmental conditions, these could take the form of noting actual disease attack or pest where the risk of attack is significant or they are likely to happen. To identify potential problems, differing growth rates could also be used.

8.5. SELECTIVE HARVESTING

At present, when the average of the whole field is ready crops are usually harvested as this simplifies the harvest process. The concept of harvesting only those crops that certain quality or quantity thresholds are involved in selective harvesting. Based on sensory perception, it can be considered to be a kind of pre-sorting. For many years in forestry, selective harvesting has been well known where according to size and quality to improve the quality of the remaining trees, certain trees are harvested.

In agriculture, examples could be to the only harvest combine grain barley below a fixed protein content that is dry enough or to harvest and select vegetables and fruits that meet certain size criteria (Figure 8.4).

Figure 8.4: Selective harvester.

Source: https://www.goodfruit.com/robotic-harvester-will-selectively-thin-and-pick-fruit/.

Improved economic returns could validate the additional sensing, as these criteria frequently attract quality premiums. Shineers et al. (2003), reported for corn harvested for biomass, benefits of the multiple-pass system rather than single-pass, as well as the benefits of asparagus (Cembali et al., 2005) and dates for selective harvesting in Iran (Abounajmi, 2004). Two criteria have to be satisfied to perform selective harvesting effectively; the capability to harvest the product of interest without injuring the remaining crop and the capability to sense the quality factor before harvest.

Most agricultural vehicles not suited for this approach because of their bigger size. Hence, for this purpose more versatile and smaller selective harvesting equipment is required. Either the harvester may have sensors mounted that can determine the crop state or before harvest, the crop can be surveyed so that the info needed about where the crop of interest is positioned. Then this ready crop can be harvested by the selective harvester while leaving the rest to dry, mature, or ripen, etc.

Otherwise, o selectively gather the entire crop from a selected area small autonomous whole crop harvesters could be used along with its transport to a stationary processing system that could sort, clean, and pack the produce. This is not a novel idea, but appraising a system that used motionless threshing machines from many years ago. On the other hand, to gather the cereal heads and send them for threshing a stripper header could be used. The harvesting process will be phased over a longer period, as selective harvesting only harvests products of the desired quality, negating the need for large equipment. It also adds value to the products, as the products are already sorted or graded before they leave the farm.

8.6. ECONOMIC SCENARIOS

The benefits and costs of potential commercial use of robotic vehicles have been compared in this section. Calculations are based on partial budgeting, where the change in cost is compared to traditional practices. Changes in initial investments change in speed, energy consumption, labor costs, control, and surveillance costs, and daily working hours are included in this model. Spraying costs saved labor costs and additional electronic devices costs, platforms for the robotic systems, and GPS-system are compared with conventional manned systems. Based on recommendations from experts and research groups, technical parameters such as capacities, dimensions, speed, and related costs. Though the economic statistics such as maintenance costs, real interest rate (5%), and period of depreciation were based on the authors' assumptions.

For the two scenarios: weeding and field scouting for weed detection, in conventional farms he model was built to project potential applications with average crop rotations and field size. Currently, at a pre-commercial development stage, these two application technologies are fairly mature.

From general economic statistics, the data for the conventional applications were taken regarding specific costs for contracting, for the Danish farm management standards. Additionally, data from researchers at the Danish Institute of Agricultural Sciences (DIAS), Bygholm, have been received, which is working on an autonomous platform with these specific applications. Autonomous Plant Inspection (API) research platform is the autonomous platform that they are using (Bak and Jakobsen, 2004).

With linear depreciation, the period of depreciation is assumed to be 10 years. The real interest rate was 5% and it was assumed that it would be essential to add some added labor time (two weeks) for making the necessary calibrations and testing for the systems.

8.7. FIELD SCOUTING FOR WEED DETECTION

Weed mapping is the process of recording the position, density (biomass), and preferably species of different weeds using machine vision. For automatic weed detection, several studies have been performed by applying different discrimination and classification techniques. Manh et al. (2001) used parametric deformable templates to segment individual weed leaves, Sokefeld et al. (2000) identified more than 20 weed species using Fourier descriptors and shape parameters, while Sogaard (2005) used Active Shape Models to identify 19 weed species. Artificial neural networks (ANNs) have also been used by many researchers to discriminate weeds (e.g., Burks et al., 2005; Granito et al., 2005) with machine vision. Other researchers have used image analysis techniques for weed discrimination using remote sensing from ground sensors (e.g., Vrindts et al., 2002) and airborne photography (e.g., Godwin and Miller, 2003).

Weed mapping is a hypothetical scenario in the sense that most farmers do not conduct systematic field scouting in their cereal fields today. Farmers either practice conventional farming with conventional spraying or they conduct organic farming with mechanical weeding. In this comparison, we assume that the alternative to autonomous weed mapping is manual weed mapping, which implies that the farmer has to register and map the weeds in the field manually with a handheld GPS.

In the weed mapping scenario, we compared autonomous field scouting for weeds in cereals with the manual detection of weeds. The autonomous system requires an API vehicle and cameras for weed detection and mapping. The DIAS has performed tests using such a vehicle for weed recognition and data from personal communications with the researchers have been used for our calculations. The API platform was initially developed by Madsen and Jakobsen (2001). Now, there is the third generation of API vehicles, further developed by Aalborg University in Denmark. This prototype has four-wheel-drive, four-wheel steering with two motors per wheel, one providing propulsion and the other steering to achieve higher mobility (Bisgaard et al., 2004).

The platform has a height clearance of 60 cm and a track width of 1 m. It is equipped with a Real-Time Kinematic Global Positioning System (RTK-GPS) and on the top of the frame; there is an operating console and an implement for the agricultural operation such as spraying devices, sensors or weeding tools. The vehicle communicates with the farm management PC for navigation, according to the computed route plan, as well as collision avoidance (Bak and Jakobsen, 2004). Based on shape recognition of weeds it is possible to create a weed map of the entire field.

The area for weed mapping and field scouting, in this scenario, is limited to 500 ha with the necessary flexibility to match large production units. Cereal crops have been focused here but it may also be helpful for other crops such as high-value crops like potatoes, sugar beet, or other horticultural crops. The lower will be the overall capacity required the shorter the time for carrying out the activity.

In weed mapping and field scouting, manual detection of weeds is compared with the robotic system. In the first year, manual scouting will take place, most of the time and in the following years; it is followed by shorter update scouting. It is estimated that about 0.7 man h/yr/ha is required in manual weed scouting (Pedersen, 2003). By using GIS and GPS systems, the weed patches are registered to produce weed maps of the separate fields. Using the API platform, autonomous field scouting has a capacity of 4.32 ha/h and a speed of 3.6 km/h, which on a 500 ha area adds up to 116 h/y for autonomous weed scouting giving 0.232 h/yr/ha (Figure 8.5).

Figure 8.5: The API platform.

Source: https://www.researchgate.net/publication/221787093_Agricultural_Robots_Applications_and_Economic_Perspectives.

Farmers would be enabled by an automated weed mapping system to generate weed maps, which will be helpful to perform the precise patch spraying with the right dose and mixture rates of herbicides where it is required. The decision support systems and the technology for the patch spraying are available and herbicide savings can change from 30% up to 75% (Heisel et al., 1999; Sogaard, 2005).

In cereals, the autonomous field scouting system costs by about 20% compared to manual weed scouting but since the system, as presented here, is only used 116 h/y, it should be possible to increase the capacity to 1000–2000 ha (Pedersen et al., 2006). Since the basis of the approximated cost of producing the platform is the costs of the autonomous platform, it might be possible that the commercial selling price will be substantially higher.

REFERENCES

1. Abounajmi, M., (2004). *Mechanization of Dates Fruit Harvesting* (pp. 1–33). ASAE Paper No. 041028. St. Joseph Mich.: ASAE.
2. Bak, T., & Jakobsen, H., (2004). Agricultural robotic platform with four wheel steering for weed detection. *Biosystems Engineering, 87*(2), 125–136.
3. Bisgaard, M., Vinther, D., & Østergaard, K. Z., (2004). *Modeling and Fault-Tolerant Control of an Autonomous Wheeled Robot* (pp. 1–43). Group Report 04gr1030a. Institute of Control Engineering, Aalborg University, Denmark.
4. Blackmore, S., Griepentrog, H. W., Fountas, S., & Gemtos, T. A., (2007). Specifications for an autonomous crop mechanization system. *Agricultural Engineering International: The CIGR Ejournal* (Vol. IX, pp. 1–30). Manuscript PM 06 032.
5. Burks, T. F., Shearer, S. A., Heath, J. R., & Donohue, K. D., (2005). Evaluation of neural network classifiers for weed species discrimination. *Biosystgems Engineering, 91*(3), 293–304.
6. Cembali, T., Folwell, J. R., Ball, T., & Clary, D. D., (2005). *Economic Comparison of Selective and Non-Selective Mechanical Harvesting of Asparagus* (pp. 1–23). ASAE Paper No. 053003. St. Joseph Mich.: ASAE.
7. Chi, Y. T., & Ling, P. P., (2004). *Fast Fruit Identification for Robotic Tomato Picker*. ASAE Paper No: 043083. St. Joseph Mich.: ASAE.
8. Christensen, L. K., Upadhyaya, S. K., Jahn, B., Slaughter, D. C., Tan, E., & Hills, D., (2005). Determining the influence of water deficiency on NPK stress discrimination in maize using spectral and spatial information. *Precision Agriculture Journal, 6*, 540–550.
9. Godwin, R. J., & Miller, P. C. H., (2003). A review of the technologies for mapping within-field variability. *Biosystems Engineering, 84*(4), 393–407.
10. Godwin, R. J., Earl, R., Taylor, J. C., Wood, G. A., Bradley, R. I., Welsh, J. P., Richards, T., et al., (2001). *Precision Farming of Cereal Crops: A Five-Year Experiment to Develop Management Guidelines* (pp. 320–328). Project report 264e. London, UK: Home Grown Cereals Authority. FAO-statistics (2005). www.fao.org (accessed on 28 May 2020).

11. Goense, D., (2003). The economics of autonomous vehicles. In: *Proceedings of the VDI-MEG Conference on Agricultural Engineering* (Vol. 1, pp. 1–22). VDI-Tatung Landtechnic, Hannover, Germany.
12. Graglia, E., (2004). Importance of herbicide concentration, number of droplets and droplet size on growth of *Solanum nigrum* L., using droplet application of glyphosphate. *XII International Colloquium on the Biology of Weeds*, (Vol. 12, pp. 1–43).
13. Granitto, P. M., Verdes, P. F., & Ceccatto, H. A., (2005). Large-scale investigation of weed seed identification by machine vision. *Computers and Electronics in Agriculture, 47*, 15–24.
14. Griepentrog, H. W., Nørremark, M., Nielsen, H., & Blackmore, B. S., (2005a). Seed mapping of sugar beet. *Precision Agriculture Journal, 6*, 2157–165.
15. Griepentrog, H. W., Nørremark, M., Nielsen, J., & Soriano, J. F., (2006). Close-to-crop thermal weed control using a CO_2 laser. In: *Proc. Proceedings of CIGR World Congress* (Vol. 1, pp. 1–33). Bonn Germany: EurAgEng, CIGR.
16. Griepentrog, H. W., Skou, P. T., Soriano, J. F., & Blackmore, B. S., (2005b). Design of a seeder to achieve highly uniform sowing patterns. In: Stafford, J., (ed.), *Proc. 5th European Conference on Precision Agriculture* (Vol. 1, pp. 675–682). Uppsala, Sweden: Wageningen Academic Publishers.
17. Hannan, W. M., & Burks, F. T., (2004). *Current Developments in Automated Citrus Harvesting* (pp. 1–22). ASAE Paper No. 043087. St. Joseph Mich.: ASAE.
18. Have, H., (2004). Effects of automation on sizes and costs of tractor and machinery. In: EurAgEng04: Proceedings of the European Conference of Agricultural Engineering (Vol. 285, pp. 1–21). Leuven, Belgium, European Society of Agricultural Engineers.
19. Have, H., Nielsen, J., Blackmore, S., & Theilby, F., (2005). Development and test of an autonomous Christmas tree weeder. In: Stafford, J. V., (ed.), *ECPA 05: Proceedings of the 5th European Conference of Precision Agriculture* (Vol. 5, pp. 629–635). Uppsala. Wageningen Academic Publishers, The Netherlands.
20. Heisel, T., Christensen, S., & Walter, A. M., (1999). Whole-field experiments with site-specific weed management. In: Stafford, J. V., (ed.), *ECPA 02: Proceedings of the Second European Conference on*

Precision Agriculture (Vol. 2, pp. 759–768). Odense, Denmark.

21. Kondo, N., Ninomiya, K., Hayashi, S., Ohta, T., & Kubota, K., (2005). *A New Challenge of Robot for Harvesting Strawberry Grown on Table Top Culture* (pp. 1–32). ASAE Paper No: 043083. St Joseph Mich.: ASAE.
22. Madsen, T. E., & Jakobsen, H. L., (2001). *Mobile Robot for Weeding* (Vol. 1, pp. 1–19). MSc. Thesis. Danish Technical University, Lynby, Denmark.
23. Manh, A. G., Rabatel, G., Assemat, L., & Aldon, M. J., (2001). Weed leaf image segmentation by deformable templates. *Journal of Agricultural Engineering Research, 80*(2), 139–146.
24. Miranda, F. R., Yoder, R. E., Wilkerson, J. B., & Odhiambo, L. O., (2005). An autonomous controller for site-specific management of fixed irrigation systems. *Computers and Electronics in Agriculture, 48*, 183–197.
25. Norremark, M., & Griepentrog, H. W., (2004). Physical methods to control weeds within crop rows. In: *EurAgEng04: Proceedings of the European Conference of Agricultural Engineering* (Vol. 405, pp. 1–44). Leuven, Belgium, European Society of Agricultural Engineers.
26. Pedersen, S. M., (2003). *Precision farming-Technology Assessment of Site-Specific Input Application in Cereals* (Vol. 1, pp. 1–32). PhD Dissertation, IPL, Danish Technical University, Lyngby, Denmark.
27. Pedersen, S. M., Fountas, S., & Blackmore, B. S., (2007). Economic potential of robots for high value crops and landscape treatment. *Precision Agriculture* (Vol. 08457–08465, pp. 1–43). Wageningen Academic Publishers.
28. Pedersen, S. M., Fountas, S., Have, H., & Blackmore, B. S., (2006). Agricultural robots-system analysis and economic feasibility. *Precision Agriculture, 7*(4), 295–308. Springer.
29. Pilarski, T., Happold, M., Pangels, H., Ollis, M., Fitzpatrick, K., & Stentz, A., (2002). The Demeter system for automated harvesting. *Autonomous Robots, 13*, 9–20.
30. Shinners, J. K., Binversie, N. B., & Savoie, P., (2003). *Whole-Plant Corn Harvesting for Biomass: Comparison of single-pass and Multiple-Pass Harvest Systems* (pp. 1–41). ASAE Paper No: 036089. St. Joseph, Mich.: ASAE.

31. Søgaard, H. T., & Lund, I., (2005). Investigation of the accuracy of a machine vision-based robotic micro spray system. In: *Proc. 5th European Conference on Precision Agriculture, 1*, 1–33.
32. Søgaard, H. T., (2005). Weed classification by active shape models. *Biosystems Engineering, 91*(3), 271–281.
33. Sökefeld, M., Gerhards, R., & Kühbauch, W., (2000). Site-specific weed control—from weed recording to herbicide application. In: *Proceedings of the 20th German Conference on Weed Biology and Weed Control* (Vol. 20, pp. 1–44). Stuttgart-Hohenheim, Germany.
34. Tothill, I. E., (2001). Biosensors developments and potential applications in the agricultural diagnosis sector. *Computers and Electronics in Agriculture, 30*(1–3), 205–218.
35. Vrindts, E., De Baerdemaeker, J., & Ramon, H., (2002). Weed detection using canopy reflection. *Precision Agriculture, 3*, 63–80.
36. Weiner, J., & Olsen, J., (2007). The influence of Triticum aestivum density, sowing pattern and nitrogen fertilization on leaf area index and its spatial variation. *Basic and Applied Ecology*, *8*(3), 252-257.
37. Wilson, J. N., (2000). Guidance of agricultural vehicles-a historical perspective. *Computers and Electronics in Agriculture, 25*, 1–9.

INDEX